GREEN ROOF CONSTRUCTION AND MAINTENANCE

McGRAW-HILL'S GREENSOURCE SERIES

Gevorkian
Solar Power in Building Design: The Engineer's Complete Design Resource

GreenSource: The Magazine of Sustainable Design
Emerald Architecture: Case Studies in Green Building

Haselbach
The Engineering Guide to LEED—New Construction: Sustainable Construction for Engineers

Luckett
Green Roof Construction and Maintenance

Melaver and Mueller (eds.)
The Green Building Bottom Line: The Real Cost of Sustainable Building

Yudelson
Green Building Through Integrated Design

About *GreenSource*

A mainstay in the green building market since 2006, *GreenSource* magazine and GreenSourceMag.com are produced by the editors of McGraw-Hill Construction, in partnership with editors at BuildingGreen, Inc., with support from the United States Green Building Council. *GreenSource* has received numerous awards, including American Business Media's 2008 Neal Award for Best Website and 2007 Neal Award for Best Start-up Publication, and FOLIO magazine's 2007 Ozzie Awards for "Best Design, New Magazine" and "Best Overall Design." Recognized for responding to the needs and demands of the profession, *GreenSource* is a leader in covering noteworthy trends in sustainable design and best practice case studies. Its award-winning content will continue to benefit key specifiers and buyers in the green design and construction industry through the books in the *GreenSource* Series.

About McGraw-Hill Construction

McGraw-Hill Construction, part of The McGraw-Hill Companies (NYSE: MHP), connects people, projects, and products across the design and construction industry. Backed by the power of Dodge, Sweets, *Engineering News-Record* (*ENR*), *Architectural Record*, *GreenSource*, *Constructor*, and regional publications, the company provides information, intelligence, tools, applications, and resources to help customers grow their businesses. McGraw-Hill Construction serves more than 1,000,000 customers within the $4.6 trillion global construction community. For more information, visit www.construction.com.

GREEN ROOF CONSTRUCTION AND MAINTENANCE

KELLY LUCKETT, LEED AP

New York   Chicago   San Francisco   Lisbon   London   Madrid
Mexico City   Milan   New Delhi   San Juan   Seoul
Singapore   Sydney   Toronto

The McGraw·Hill Companies

Cataloging-in-Publication Data is on file with the Library of Congress

Copyright © 2009 by The McGraw-Hill Companies, Inc. All rights reserved. Printed in the United States of America. Except as permitted under the United States Copyright Act of 1976, no part of this publication may be reproduced or distributed in any form or by any means, or stored in a data base or retrieval system, without the prior written permission of the publisher.

1 2 3 4 5 6 7 8 9 0   DOC/DOC   0 1 5 4 3 2 1 0 9

ISBN 978-0-07-160880-0
MHID 0-07-160880-X

Sponsoring Editor: Joy Bramble Oehlkers
Editing Supervisor: Stephen M. Smith
Production Supervisor: Pamela A. Pelton
Developmental Editor: Rebecca Behrens
Acquisitions Coordinator: Michael Mulcahy
Project Manager: Ekta Dixit, International Typesetting and Composition
Copy Editor: Ragini Pandey, International Typesetting and Composition
Proofreader: Megha Roy Choudhary
Indexer: Robert Swanson
Art Director, Cover: Jeff Weeks
Composition: International Typesetting and Composition

Printed and bound by RR Donnelley.

McGraw-Hill books are available at special quantity discounts to use as premiums and sales promotions, or for use in corporate training programs. To contact a representative, please e-mail us at bulksales@mcgraw-hill.com.

The pages within this book were printed on acid-free paper containing 100% postconsumer fiber.

About the Author

Kelly Luckett, LEED AP, is president of Saint Louis Metalworks, a provider of sheet metal roof accessories and metal roofing. In 2004, his company launched the Green Roof Blocks product line and has since installed more than 50 green roofs on such buildings as universities, convention centers, businesses, and private residences. He has published newsletter and magazine articles and is a member of Green Roofs for Healthy Cities, a rapidly growing not-for-profit industry association working to promote green roofs throughout North America.

CONTENTS

PREFACE

Green roofs have been increasing in popularity in the United States for the last 10 years and in Europe over the last 40 years. The green roof concept represents a truly rare occurrence in modern economics: the creation of new markets for existing goods and services.

The irresistible lure of new business opportunity has drawn companies from across a broad range of industries. As these players step up to stake their claim in the green roof game, they are faced with the reality that there are very few absolutes and the risk level is quite high. Further exciting the marketplace are emerging standards from various regulating and authoritative agencies. American Standards and Testing Methods, Factory Mutual, and National Roofing Contractors Association, to name just a few, either have published or intend to publish guidelines for green roof construction. Challenged with producing a compilation of information gathered from a variety of sources, it is easy to produce an overly general set of guidelines.

Obviously, there is no single set of green roof instructions that applies to every situation. The green roof construction guidelines on the following pages are no exception. However, the intent of this book is to cut through some of the proprietary overkill and overly general information, to offer some practical information gathered through real-world installations and independent research.

Kelly Luckett, LEED AP

GREEN ROOF CONSTRUCTION AND MAINTENANCE

Design

Green roof design begins with developing a complete vision of the green roof and the building it adorns. The design process begins with asking a series of questions that will help define the various aspects of the green roof being designed. Why build a green roof on this particular project? More specifically, what is the purpose of this green roof? Will people gather on this roof to enjoy and interact with nature? If so, there are specific structural requirements that must be met to support this type of usage. Life safety requirements may mandate the construction of railings and roof edge setbacks that ensure the safety of those using this space. If people are to enjoy the green roof, they will need a safe and convenient means of getting up there. Direct access doors from sections of the building serviced by elevators will allow more people to gain access than requiring visitors to traverse stairs or crawl through windows. (See Fig. 1.1.)

Figure 1.1   Access through a door in the penthouse and safety barricades at the roof edges make this an inviting gathering spot for the building occupants.

Perhaps the green roof will not be a place for people to gather, but rather an aesthetically pleasing alternative rooftop visible from key areas within the building. Yet again, maybe the green roof will not be seen by anyone. Perhaps the green roof is part of a plan to reduce storm water runoff, reduce energy consumption, gain LEED certification, or do all of the above. We may sometimes have a clear vision from the beginning of how the green roof should look and what purpose it will serve. However, more often the vision will evolve during the design and construction process. For example, one may begin with a vision of a green roof with growth media depths capable of supporting tall plants and native grasses only to run up against structural loading limitations that force him or her to alter the plans to a shallower design with succulents and simple ground covers. (See Fig. 1.2.)

Figure 1.2   Simple plant selections comprising low-lying ground cover and succulents can thrive in shallow growth media depths on green roofs desiring minimum weight.

Green roof design often requires balancing desire, need, and finances. Establishing a green roof can be a simultaneously rewarding and exasperating experience. One can take an otherwise drab and unappealing area atop a building and transform it into a remarkable green space. One may also spend several months working on a project only to have the green roof eliminated due to budgetary constraints. In order to give a green roof project every chance to succeed, it is important to clearly identify the intent and limitations specific to the project. The first section of this book will discuss design options, benefits, and ramifications. Designing a green roof is not necessarily a linear process. Although it is best to attempt to address issues in a logical order, one must often revisit earlier portions of the design to tweak, refine, and on occasion, scrap and start over. The goal is to design a green roof that will come to fruition,

Figure 1.3   This Washington, D.C., green roof has incorporated a play area with rubber mulch to enhance the interaction with rooftop visitors.

meet the intent of the design, and become a source of pride for its owner. (See Fig. 1.3.)

Why Create This Green Roof?

Why is the owner of this structure interested in building this particular green roof? The answers to this question will give those involved in the project some direction as they develop their vision of the green roof project. To obtain the answers needed, one must ask some more specific questions about the green roof he or she wants to build. It's best to start with the desired use of the rooftop space.

Will people gather and congregate on the green roof? If the answer is yes, then the green roof is likely a rooftop garden for the enjoyment of visitors. If the answer is no, the green roof will be designed to meet visual, environmental, and regulatory goals.

Many design considerations will need to be addressed based on building a rooftop garden versus a green roof.

EGRESS AND ACCESS

Rooftop garden   How will visitors get to the rooftop garden? Exterior doors that allow access to the rooftop either through an adjacent section of the building or through a penthouse located on the roof will be required. Such doors become subject to the same security considerations as any other means of entry into our building and will need to be integrated into the property's security policy. Access to the rooftop door could mean facilitating access to certain sections of the building and may require some consideration of possible disruption of building activities. Often stairways are the only means of access to the rooftop. New construction projects may be required to meet the Americans with Disabilities Act of 1990 (ADA)* while green roof projects on existing buildings may be exempt. Complying with ADA could include restrictions on thresholds, doorway widths, and door hardware, as well as require the use of an elevator or ramp. Consult with the local building code enforcement agency to determine how ADA requirements will affect the green roof design. (See Fig. 1.4.)

Green roof   Even if the green roof is not designed to be a public gathering space, access to the rooftop still needs to be addressed. Access will be necessary for the construction and routine maintenance of the green roof; those issues will be discussed in detail further into this book. Many rooftops are routinely designed with some form of access, whether by means of a roof hatch or an exterior door. In the absence of an exterior door through an adjacent section of the building or through a penthouse located on the roof, a roof hatch allows rooftop access through the roof surface from the level just below the rooftop. Roof hatches can be easily designed into a new construction project, but can also be easily added to an existing building. Typically, roof hatches are installed with permanent access ladders that are concealed in building storage closets

---

*United States Public Law 101-336, 104 Stat. 327 (July 26, 1990), codified at 42 U.S.C. § 12101 et seq.

Figure 1.4   This new St. Louis Community College building uses a combination of ramps and elevators to ensure access to all areas of the building for handicapped occupants, including access to this green roof.

and maintenance rooms. Small projects may not accommodate interior access to a roof hatch. It may be necessary to plan roof access using a ladder. Design considerations for ladder safety may include anchors to tie off the top of the ladder or a level area on the ground for setting the base of the ladder. What's important here is to note the necessity for access to the green roof even if the design is not for a rooftop garden. (See Fig. 1.5.)

STRUCTURAL SUPPORT

Rooftop garden   Having determined that the rooftop garden will be a place for people to gather and having provided access, one must ensure that the roof structure has the necessary structural capacity to support rooftop activity. Building codes may vary, so it is important to determine the local requirement for live

Figure 1.5 Access to the green roof on a Bank of America branch bank is provided by a roof hatch located over a storage room. A steel ladder is permanently mounted to the wall allowing for access to the roof level through the interior of the building.

loads and dead loads, and to understand how the green roof being built relates to each weight requirement. The entire green roof assembly, including plants and the water required to saturate the growth media, is considered part of the dead load of the structure. Water in excess of that which saturates the growth media, snow, and people visiting the green roof are all considered part of the live load of the structure. One must formulate a preliminary idea of what type of plants are desired and the proper growth media depth required to support them. Saturated weight data should be available from the manufacturers of the intended green roof components. Typical rooftop gardens incorporate varying growth media depths and planters to support various plant choices. This will require calculations of the point loading of these various areas to determine structural requirements. Evaluating loading requirements and upgrading the structure to support the green roof is easiest and most economical in the design phase of the construction of the building. Evaluating the structural capacity and making upgrades to an existing structure is significantly more difficult and more expensive. Many retrofit green roof plans die at this stage due to inadequate structural capacity and the prohibitive cost of upgrades. While there are some creative strategies of employing irrigation systems to reduce growth media depths in order to reduce dead loading, live load requirements could mean abandoning public accessibility to the rooftop garden and opting to design a simpler, extensive green roof. (See Fig. 1.6.)

Green roof When the green roof will not be a public gathering space, the live load structural requirements for the green roof are less complicated. Once the load requirements of the local building code have been determined, one must calculate the saturated weight of the green roof system to determine if structural upgrades will be necessary. Again, this is going to require some idea of the type of plants intended to grow on the green roof and the growth media depth required to support them. Typically the entire green roof will have a uniform dead

Figure 1.6   The Olsen Garden on the rooftop at the St. Louis Children's Hospital includes elaborate water features, shade-providing trees, and artistic sculptures. Heavier features can be strategically located over structurally supported areas of the roof.

load based on the saturated weight of the green roof assembly, though one may consider positioning planters or mounded growth media over structural support members to incorporate some strategically located deeper growth media for larger showcase plantings. The plant palette is significantly expanded by increasing the growth media depth. As increased depth results in increased weight, there are often tradeoffs that balance structural cost with plant selection. Once the dead load of the green roof has been determined, a new structure can be designed with the required capacity. For an existing structure, one must begin by determining the structural capacity and design within those parameters. Irrigation systems have been successfully used to reduce growth media depth, and thus the weight of the green roof system, for projects that would have otherwise required costly structural upgrades. For example, the green roof on the Ford Rouge Dearborn Truck Plant thrives in less than 3 inches of growth media and is sustained during periods of drought by the strategic use of supplemental irrigation. (See Fig. 1.7.)

Figure 1.7 Supplemental irrigation can allow a reduction of growth media depth to reduce the weight of the green roof system. This green roof on the Ford Truck Plant is less than 3 inches deep.

LIFE SAFETY

Rooftop garden   Once visitors arrive at the rooftop garden, their safety must be ensured. This may require perimeter railings or fencing to keep visitors from falling off the roof. However, it could also include skid-resistant walking surfaces, ample lighting, GFI-protected outlets, handrails at steps, and well-defined boundaries separating accessible areas from areas containing rooftop equipment. Secure storage areas will be necessary to keep maintenance tools and chemicals safely stored. Designated hours of operation should be considered to allow routine maintenance to be conducted when visitors to the roof garden are not present. Some consideration should be given to activities conducted in adjacent buildings to ensure the safety of our visitors from falling objects from windows, scaffolding, fire escapes, balconies, and the like. (See Fig. 1.8.)

Figure 1.8   This St. Louis rooftop garden uses the green roof plants to define areas designated for rooftop visitor access by providing a 12-foot-wide buffer between the observation deck and the roof edges.

Green roof Even if a green roof is not designed to accommodate visitors, there are some safety issues/features that must be incorporated into the design. Fall protection will be required for anyone working within 10 feet of the roof edge. This includes anyone engaged in the initial construction of the green roof as well as those conducting routine maintenance. The most cost-effective manner to address fall protection is to incorporate provisions into the design. The Occupational Safety and Health Administration* (OSHA) requires anchor points to be capable of supporting 5000 pounds per worker anchored to that point. For example, a cable running between two anchor points to which three workers are attached would be required to be capable of supporting 15,000 pounds at each point of anchorage. Rooftops with parapet walls at least 40 inches tall do not require additional fall protection measures. Small projects that will require a ladder for access will need to have an anchor point to secure the ladder. As with fall protection, it is easiest and most cost effective to integrate the anchor point for a ladder into the design. (See Fig. 1.9.)

Where Is the Green Roof Being Built?

At this point, the general scope of the project has been determined and some weight and access issues that will affect the final design have been considered. However, the focus has been limited to the rooftop; next, it will be expanded to include the area around the building. This section will look at the project location and the important elements that will impact the green roof. In this important phase of design, one must consider the conditions that will shape the plant selection palette. Here the idea of the color and texture of the rooftop space is fleshed out. The orientation of the roof area will be considered first.

---

*U.S. Department of Labor, Occupational Safety and Health Administration, Fall protection systems criteria and practices—1926.502

Figure 1.9   The use of personal fall protection devices such as body harnesses and retractable lanyards keep workers safe from falling from the rooftop. Other safety practices include the use of proper clothing, work boots, hard hats, and safety glasses.

ELEVATION

Hoisting   The height of the building section dictates several aspects of the green roof design. It has been stated that good green roof construction is all in the material handling. The volume and weight of the green roof material, especially growth media, makes roof loading a challenge. The taller the building, the more expensive it will be to get material to the rooftop. The following categories simplify consideration of height for material hoisting: rooftops below 20 feet, rooftops between 20 and 120 feet, and rooftops higher than 120 feet. Rooftops below 20 feet may be stocked using extending-boom forklifts and common roof-loading equipment. This equipment is typically leased and operated by onsite construction personnel, affording flexibility in the scheduling of material deliveries. Rooftops between 20 and

120 feet typically require the use of a crane to hoist material to the rooftop. Conducting operations at this height requires a great deal of skill and coordination. Using a crane to hoist material requires experience in rigging loads for hoisting, the use of hand signals and radio equipment to communicate with the crane operator, understanding of point loading capacity of the roof deck, and safety practices and regulations for rooftop activities and other hoisting procedures. The typical hoisting operation is conducted by six or more highly paid tradesmen. Coordination of material delivery times and crane availability is critical to avoid paying workers to wait on the arrival of hoisting equipment. Projects exceeding 120 feet in height are often stocked using a tower crane or an elevator. Access to tower cranes is regulated through the strict scheduling of time slots. The hoisting plan often includes rehearsals to eliminate all unnecessary steps and to streamline the process. Time slots are allocated to various trade groups working on the project. The crane must be available for the next scheduled trade group or else every subsequent trade group will be delayed. While not as rigidly policed, construction elevators on construction projects and freight elevators on retrofit projects have similar scheduling constraints. Some consideration must be given to moving material through the building using the elevator in the least disruptive manner possible. This may require stocking materials on weekends or after hours adding the cost of premium pay to the green roof project. Alternatively, some projects use large blowers to transport growth media and plant material to the roof through a hose. This method allows growth media to be quickly distributed over large areas. Taller elevations make this process more difficult and add cost proportionately as the height increases. (See Fig. 1.10.)

Wind Winds are often greater at higher elevations. Consideration of winds blowing across the rooftop will affect the plant selection and placement. Taller, upright plants catch wind and may need to be situated away from the roof edge where winds are stronger. Supplemental anchoring may be necessary to allow these taller plants a chance to establish roots capable of withstanding wind loads. The perimeter of the roof is affected by a phenomenon known as *wind vortex*, where wind travels up the wall of the building and creates negative pressure

Figure 1.10   Access to the rooftop varies and requires a project-specific plan for transporting materials to the rooftop.

Figure 1.10   (*Continued*)

at the roof surface as it swirls along the roof edge.* It may be necessary to incorporate "no plant zones" in these areas. Often concrete pavers are used to add additional weight to counter wind uplift forces along the roof perimeter. Even on the interior regions of the roof, strong winds can wreak havoc on the green roof surface. Well-rooted and established plants help hold the growth media to prevent scouring. The use of wind blankets may be necessary to offer protection against scouring until plants can be established. *Wind blankets* are geo-textile materials that cover the green roof, shielding the surface of the growth media from the wind. The wind blanket is anchored in place and the plants are propagated through small openings cut in the wind blanket material. The wind blanket is designed of organic material to decompose gradually as the plants mature and cover the roof surface; this provides the added benefit of supplying organic nutrients to the growth media. High wind loading on a green roof project will require frequent inspections in order to correct small problems before they turn into big problems. (See Fig. 1.11.)

ADJACENT STRUCTURES AND RESULTING MICROCLIMATES

Shadowing   Taller structures in close proximity to the green roof may cast shadows over the green roof. Depending on the orientation, shadows may fall over the roof surface at various times throughout the day as well as throughout the year. Care must be taken to match the sunlight requirement of the selected green roof plant species with the sunlight available on the rooftop. Partial shading presents a growing challenge, but can also be used to one's advantage. Plants that are less tolerant of heat can be positioned to take advantage of the morning sunlight while afternoon shading provides relief from the heat. Green roofs in constant shade require careful plant selection. Some plant species may have a distinctly different appearance when fully shaded than when exposed to direct sunlight. Some species will not tolerate winter shading in northern climates. It is best to embark on shaded green roof projects with realistic expectations regarding the trial and error necessary to develop a plant scheme that will thrive on the chosen rooftop. (See Fig. 1.12.)

---

*ANSI/SPRI RP-4 2002, Wind Design Standard for Ballasted Single-Ply Roofing Systems

Figure 1.11   The wind blanket is designed of organic material to decompose gradually as the plants mature and cover the roof surface.

Figure 1.12   Consideration must be given to shading from an adjacent structure. Shading may influence plant selection and planting density and may require ongoing adjustments.

Reflection   Adjacent vertical walls constructed of glass or reflective metal cladding reflect and amplify the intensity of sunlight. Sunrays bouncing off these surfaces can quickly dry out the growth media and dehydrate plants. These areas require extremely drought- and heat-tolerant plant species and may require more frequent irrigation. Increasing the growth media depth will help to buffer the heat gain and retain more water to help keep the plants hydrated. The radiant heat in these areas may also extend the plant growth period in northern climates by warming the roof surface. (See Fig. 1.13.)

Access   Adjacent building sections can both enhance and inhibit access to the rooftop on which the green roof will be built. Doorways and windows through adjacent building sections provide easy access to the rooftop. However, adjacent building sections may prohibit crane and forklift access to the rooftop, requiring material to be conveyed through the building or over the top of adjacent roofs. This requires double handling of material and can significantly increase the cost of the green roof. Modular green roof systems are particularly attractive for

Figure 1.13   Adjacent building windows can affect the green roof plants in both a positive and negative manner. Consideration of varying microclimates within a single rooftop may require employing area-specific plant and irrigation strategies.

Figure 1.14   Roof sections that are surrounded by taller building sections may require green roof materials to be conveyed through the interior of the building.

these situations as the growth media and plants are contained in individual modules. (See Fig. 1.14.)

Exposure   Taller adjacent building sections can dramatically reduce the wind and extreme weather exposure for a green roof. Green roofs oriented downwind of prevailing weather patterns will experience less wind-driven rain and snow. However, upwind orientation will result in greater snow drifts against the taller structure and greater accumulation of water along the base of the adjacent wall. (See Fig. 1.15.)

Taller adjacent building sections can also provide visibility of the green roof from the interior of the building. Planting strategies for these green roofs may include evergreen species that provide vibrant foliage during the winter months. Flowering annuals can be used for adding spring and summertime color. Green roofs that are visible from the interior of the building become an amenity allowing interaction with the occupants and adding value to the property. (See Fig. 1.16.)

Figure 1.15   Taller adjacent building sections can dramatically reduce the wind and extreme weather exposure for a green roof.

Figure 1.16   Views of a green roof from overlooking windows will make the green roof an important amenity for the property, possibly increasing leasing rates and resale value.

Rain water harvesting Taller adjacent buildings afford a unique opportunity to capture rainwater from higher elevations to use for the irrigation of a green roof. Strategic location of gutters and downspouts can allow the use of rain barrels and cisterns to store rainfall while diverting storm water runoff from overburdened storm water systems. The capacity of the rain barrels or cisterns, combined with the retention capacity of the growth media, can play an important role in the overall project storm-water management plan. It is necessary to consider the point loading of these rainwater storage strategies to ensure that there is adequate structural capacity to support the weight of the collection vessel. Ongoing research is studying the collection of condensation water from rooftop air-conditioning equipment that may expand the water-harvesting strategy beyond that of only rainfall. As supplemental irrigation dramatically expands the plant select options and may also permit a reduction of growth media depth to meet structural limitations, rainwater harvesting is becoming an integral component of green roof design. (See Figs. 1.17 and 1.18.)

Figure 1.17 Strategic location of gutters and downspouts can allow the use of rain barrels and cisterns to store rainfall while diverting storm water runoff from overburdened storm water systems.

Figure 1.18 Water harvested from upper roof levels can be used to irrigate green roof plants during periods of drought.

Rooftop equipment Heating, air-conditioning, and ventilation (HVAC) equipment located on the rooftop can impact the green roof design and vice versa. Discharge from exhaust fans can contain fumes that are harmful to the green roof plants. For example, kitchen exhaust fans often require grease traps, which contain fat drippings that are harmful to plants and roof membranes. Grease traps typically require routine maintenance so it is important to plan for providing access to this equipment as well as for larger HVAC equipment.

Figure 1.19 Consideration of the effect air-handling equipment will have on green roof plants may include using setbacks or identifying and selecting plant species suited for a particular microclimate.

While building exhaust may help to warm an area of the green roof during cold weather, helping to extend the growing period of the plants near the vent, green roof plants can significantly lower the ambient air temperature near the rooftop in summer months. This can result in much cooler air being drawn into the intake of the building ventilation systems, allowing the cooling system to operate more efficiently and perhaps even enabling some downsizing of the equipment. (See Fig. 1.19.)

Solar equipment Solar and photovoltaic equipment is often located on rooftops as part of a sustainable design and construction strategy. This equipment may require additional roof penetrations that must be sealed into the roofing or waterproofing. Setting plants back from these areas will allow for routine inspection of the flashings and for required repairs or

Figure 1.20   **Courtesy of Alternative Energy Matters.**

maintenance to be conducted easily. The panels often cast shade over the roof and should be factored into plant selection for these areas. Studies have shown that green roofs' abilities to lower the ambient air temperatures near the roof surface enhance the efficiency of photovoltaic operation.* (See Fig. 1.20.)

REGIONAL

Climate   The regional climate will largely dictate the particular plant species selected for a green roof; the two primary influences are temperature and precipitation. The horticulture industry uses hardiness zone categories to label the regions to which plants are suited. Hardiness zones divide the continental United States into bands based on the historic winter low temperatures. Zone 10, the southernmost band, has the warmest average winter temperatures, while zone 2, the northernmost band, has the coolest average winter temperatures. (See Fig. 1.21.)

---

*Proceedings: Rio 02, ed. by Stefan Krauter

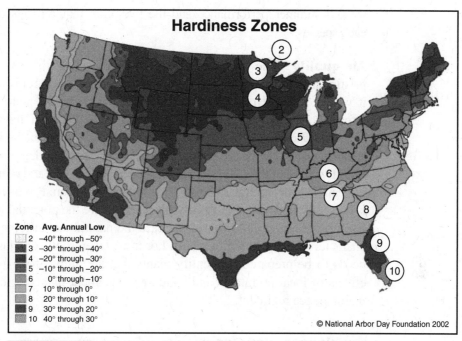

Hardiness Zones

Zone	Avg. Annual Low
2	−40° through −50°
3	−30° through −40°
4	−20° through −30°
5	−10° through −20°
6	0° through −10°
7	10° through 0°
8	20° through 10°
9	30° through 20°
10	40° through 30°

© National Arbor Day Foundation 2002

Figure 1.21 **Winter hardiness zones.**

Without getting into the global warming debate here, it is notable that the hardiness zones have moved upward one bandwidth in recent years due to warmer winters. The summers, however, have grown warmer in recent years as well. Some in the horticulture industry have begun to develop summer hardiness zones to identify heat and drought tolerances of plant species. Discussions of heat tolerance almost always accompany mention of drought tolerance. High heat exacerbates the effect of drought on green roof plants, as plants can survive much longer without rainfall in cooler temperatures. There are several resources available for consultation when selecting plants for a green roof, including the following websites: www.greenroofplants.com, www.greenroofs.org, and www.greenroof.com. When considering plant selections, look for both the winter and summer performance and the hydration requirements of each species.

Air quality The term air quality conjures either visions of

ter and summer performance and the hydration requirements of each species.

Air quality   The term air quality conjures either visions of pristine air and cobalt blue skies or smog so thick one could cut it with a knife. However, many times the air quality factors affecting green roof plants are somewhere in between those extremes. Acid rain from neighboring industrial smokestacks is seldom visible while trace contaminants can be found in the runoff water and the plant tissue. Nitrates in polluted air and rainfall can provide enough nitrogen for adequate nutrition. Research is presently underway to identify plants that have the potential to fix atmospheric pollutants and increase the natural air scrubbing characteristic of green roof plants. For most green roofs, one just needs to be prepared to identify plants that are ailing from the effects of local pollutants and include replacement plants in the maintenance regiment.

Regulations and incentives   While many projects are required to go through a permitting and inspection process, green roofs are so new in the United States that code enforcement officials lack the standards to regulate construction methods. Several agencies are currently developing standards: American Standards and Testing Methods (ASTM), National Roofing Contractors Association (NRCA), Single-Ply Roofing Industry (SPRI), and Factory Mutual are all working on construction standards or testing methods for green roof components. Developing standards is a tall order because green roof construction varies dramatically from system to system and from project to project. The industry has identified the basic components of green roof construction, but wind uplift and fire rating remains somewhat of an enigma. The Single-Ply Roofing Industry and Green Roofs for Healthy Cities are jointly sponsoring the development of prescriptive standards detailing green roof construction methods developed to address wind uplift and fire exposure concerns. Once completed, the RP-14 2007 Wind Design Standard and the VF-1 2007 Fire Design Standard will be submitted to the International Code Council for approval to be included in the 2009 publishing of the International Building Code.

Figure 1.22 This Washington, D.C., green roof is part of the storm water management strategy employed by the developer to meet the requirements of the local storm water management regulations.

Figure 1.23 For many Washington, D.C., development projects, green roofs are less expensive than employing underground cisterns to manage storm water runoff from rooftops.

Some cities and municipalities are becoming leaders in terms of green roof awareness, knowledge, and incentives. Washington, D.C., is accepting green roofs as part of the development storm water management plan instead of the storage systems typically designed into construction projects. (See Figs. 1.22 and 1.23.) For many developers, the green roof is a more cost-effective alternative for meeting the stringent storm water management regulations. Portland, OR, has a density bonus that awards additional floor space for projects adding green roof space. Developers in Chicago, Illinois, are mandated to include some green roof space on their projects; the mandate increases the required green roof space proportionate to government funding used for the project. Increasingly, communities are adopting impervious surface fees to fund storm water treatment based on the amount of hard pavement and rooftop present on each property. The green roof industry is actively lobbying these agencies to allow deductions for green roof space. Unfortunately, the fee structures currently do not cover the cost of a green roof. In contrast, European communities have begun to separate the cost of storm water treatment from sanitary sewage treatment and property owners are assessed fees based on storm water contributory

Figure 1.24   This green roof project on McCormick Place Convention Center in Chicago, Illinois, was installed utilizing a composite crew made up of members of the Laborers Union, the Roofers Union, and the Sheet Metal Workers Union.

surface area of each property. This equitable means of assessing treatment fees lowers the sanitary sewage treatment fees for most of the population, passes the actual cost of storm water treatment to those property owners responsible for the runoff, and provides incentives for storm water reduction strategies like green roofs and permeable pavement.

Organized labor and prevailing wage Organized labor unions dominate the construction industry in many regions of the United States. It is important to consider regional labor rates for construction of a green roof and be aware of the factors that affect the construction cost from one region to the next. The Department of Labor determines the prevailing wages to be paid to workers of various construction trades on federally funded construction projects. Likewise, individual states also determine the prevailing wages for construction trades in the various regions of their state. Privately funded projects may not be required to meet prevailing wage mandates. It is critical to identify when these requirements apply to green roof projects and make certain that members of the construction team for which the requirements apply are aware of their compliance responsibilities. (See Fig. 1.24.)

2

Structural

Design

N ow the purpose, the function, and many factors that will impact the design of the green roof project have been determined. However, before the vision of a green roof can be completed, green roof designers must determine the structural requirements of the green roof and the structural capacity of the rooftop. This process will require the involvement of a qualified structural engineer. The structure for new construction projects can be engineered to support the envisioned green roof. However, green roofs for existing buildings will typically need to be designed based on the limitations of the structure. It is easiest to get a structural evaluation of an existing rooftop if one has access to the construction blueprints. They are rarely available for older buildings. On occasion, however, blueprints may be available from the local public works department's archives. If not, an engineer needs to conduct a structural analysis of the roof. Here lies the most difficult phase of green roof design: the marriage of structural capacity with the green roof vision. Many green roof projects die here; those projects that survive are often the result of considerable compromise. The process often requires juggling back and forth between weight and structure to find the green roof construction strategies that meet the predesign intent and the structural limitations.

Until now, the focus has been on conceptualizing the characteristics of a green roof. Now it's time to move beyond the conceptual to identify the specific components of the green roof system to assess the weight of each.

The following sections discuss various green roof components, their functions in the green roof system, and the role water plays in structural loading. The dry weight and the saturated weight of some of these components will be discussed. Calculating the loading of the structure based on the weight of the green roof system at the point of saturation is required. The green roof system, fully saturated with water, is considered part of the dead load of the roof structure.* Excess water flowing throughout the saturated green roof system is considered part of the live load. Snow is also considered part of the live load. The engineer will calculate the structural dead load, including the green roof components, as well as the anticipated live load. If the project is a new construction, the engineer will design the structure to support the total loading. If the green roof project is to be on top of an existing building, the engineer will either approve the green roof design for the structure or otherwise prescribe any necessary structural enhancements. If the structural requirements of the system are above the capacity of the structure or drive the cost of the new structure beyond the budget, the green roof must be redesigned in order to reduce the weight. If the weight and budget issues cannot be reconciled, this particular green roof project cannot be built.

The following list provides a general overview of available green roof components. These components perform a variety of functions and some are better suited under certain conditions. During the component selection process, many green roofs are overdesigned, driving up the cost of the green roof.

---

*ASTM E 2397-05 Standard Practice for Determination of Dead Loads and Live Loads associated with Green Roof Systems

While it is a natural tendency to design in every possible safeguard, doing so typically results in exceeding the project budget. The task at hand is to separate necessity from overkill and design a successful green roof that can become a reality.

Roofing Insulation

Roof insulation is often referred to as rigid insulation because insulation materials are formed into sheets or boards that are laid flat over the roof surface. Sometimes the insulation boards are positioned below the roofing or waterproofing materials and sometimes positioned above the roofing or waterproofing materials. Roof insulation placed over the roofing or waterproofing material would need to be made from materials capable of being exposed to moisture. Additionally, some insulation materials cannot withstand exposure to high temperatures. These require the installation of an additional protective layer of roof insulation referred to as cover boards, made from materials capable of withstanding exposure to high temperatures, when used in roofing systems that are applied as heated liquids like hot bitumen, coal tar pitch, and hot liquid rubber. The following section discusses the characteristics of some common roof insulations and applications for which they are suited.

POLYISOCYANURATE

- Primary roofing insulation
- Must be kept dry
- Installed below the roofing membrane
- May have single-ply membranes directly adhered
- Must have a cover board for hot asphalt or hot rubber applications
- Weight per inch of thickness: 0.2 to 0.3 lb/ft^2
- R-factor per inch of thickness: R-6

EXTRUDED POLYSTYRENE

- Primary roofing insulation
- Impervious to water absorption
- May be installed above or below the roofing membrane

- Must have cover board for fully adhered single-ply membranes and hot asphalt and hot rubber applications
- Weight per inch of thickness: 0.25 lb/ft^2
- R-factor per inch of thickness: R-5

EXPANDED POLYSTYRENE

- Primary roof insulation
- Must be kept dry
- Installed below the roofing membrane
- Must have cover board for fully adhered single-ply membranes and hot asphalt and hot rubber applications
- Weight per inch of thickness: 0.1 to 0.2 lb/ft^2
- R-factor per inch of thickness: R-3.5

FESCO BOARD

- Primary roof insulation or cover board for foam insulations
- Must be kept dry
- Installed below the roofing membrane
- Must be adhered or loose-laid when used beneath green roofs as mechanical fasteners
- May damage the roofing membrane when the weight of the green roof is applied over the fastener locations
- Weight per inch of thickness: 0.77 lb/ft^2
- R-factor per inch of thickness: R-2.78

Roofing Membranes

EPDM

- Most commonly used membrane
- Low cost
- Large sheet size minimizes seams
- Excellent durability and root resistance
- Poor chemical and oil resistance makes EPDM a poor choice for restaurants and rooftops with exhaust hoods ventilating airborne oils
- Common thicknesses: 45 mil (0.29 lb/ft^2), 60 mil (0.40 lb/ft^2), and 90 mil (0.63 lb/ft^2)

TPO

- Increasingly popular membrane
- Reflective white surface
- Heat-welded seams
- Excellent durability and root resistance
- Good chemical and oil resistance
- The expense of heat-welding equipment can limit the number of qualified contractors, reducing the competition, and increasing the cost of the project
- Common thicknesses: 45 mil (0.232 lb/ft^2), 60 mil (0.314 lb/ft^2), and 80 mil (0.42 lb/ft^2)

PVC

- Reflective white surface
- Heat-welded seams
- Excellent durability and root resistance
- Excellent chemical and oil resistance
- The expense of heat-welding equipment can limit the number of qualified contractors, reducing the competition, and increasing the cost of the project
- Common thicknesses 45 mil (0.232 lb/ft^2), 60 mil (0.314 lb/ft^2), and 80 mil (0.42 lb/ft^2)

BUILT-UP ROOFING (BUR)

- Commonly used roofing strategy
- Often surfaced with pea gravel
- Low cost
- Poor root resistance requires the use of a root barrier to prevent plant roots from growing into the asphalt surface
- Poor chemical and oil resistance makes BUR a poor choice for restaurants and rooftops with exhaust hoods ventilating airborne oils
- Common thicknesses: 2 to 3 lb/ft^2 (add 4 lb for gravel surfacing)

MODIFIED BITUMEN

- Commonly used roofing membrane as cap sheet for built-up roofing systems
- Available in torch down (APP) and adhered (SBS) formulations
- Low cost

■ Poor root resistance requires the use of a root barrier to prevent plant roots from growing into the asphalt surface

■ Poor chemical and oil resistance makes modified bitumen a poor choice for restaurants and rooftops with exhaust hoods ventilating airborne oils

■ Common thickness: 1 to 1.75 lb/ft^2

LIQUID-APPLIED MEMBRANE

■ Increasingly popular waterproofing strategy for green roofs

■ Available in hot rubber-modified asphalt formulations and synthetic liquid membrane formulations

■ Excellent for monolithic concrete substrates

■ Poor root resistance requires the use of a root barrier to prevent plant roots from growing into the asphalt surface

■ Poor chemical and oil resistance makes liquid-applied membrane a poor choice for restaurants and rooftops with exhaust hoods ventilating airborne oils

■ Common thickness: 0.75 to 1.5 lb/ft^2

METAL ROOFING

Although green roofs over metal roofing are rare in the United States, companies in Europe have found metal roofing to be a long-lasting and lightweight roofing material, well-suited for green roofs. While initial cost is considerably higher than other roofing materials, life spans exceeding 100 years make metal roofing an attractive waterproofing strategy for institutions and government buildings. The most common thickness is 1 to 1.5 lb/ft^2.

Protection Materials

GYPSUM-BASED COVER BOARDS (DENS DECK OR DURABOARD)

There are several manufacturers of this type of product. This material must be kept dry, and is installed below the roofing membrane. This product serves to protect the insulation from heat and chemical attack during adhered and hot-applied roofing membrane installation. Further, the cover board helps to distribute point loading of rooftop traffic over a greater area, reducing the risk of damage to the insulation during green roof construction. It must be adhered or loose-laid when used beneath green roofs because mechanical fasteners may damage the roofing membrane when the weight of the green roof is applied over the fastener locations. The three most common thicknesses are 1/4-inch (1.1 lb/ft^2), and 1/2-inch (1.95 lb/ft^2), and 3/4-inch (2.5 lb/ft^2).

FESCO BOARD (WOOD FIBER)

This product must be kept dry and is installed below the roofing membrane. It serves to protect the insulation from heat and chemical attack during adhered and hot-applied roofing membrane installation. This material is not as dense as the gypsum products and thus has less point-loading durability, but produces a roof surface significantly more durable than that of insulation board alone. It must be adhered or loose-laid when used beneath green roofs as mechanical fasteners may damage the roofing membrane when the weight of the green roof is applied over the fastener locations. The most common thickness is 1/2 inch (0.46 lb/ft^2).

EXTRUDED POLYSTYRENE

This material is sometimes referred to as DOW Board and is often used as primary roof insulation. However, it is impervious to water, making it well suited to serve as both insulation and protection board by installing it above the waterproofing material. There is a drawback to using the insulation as the protection board: the weight of the green roof is used to hold the insulation in place. Therefore, the insulation must immediately be covered by the green roof material; this requires careful coordination to minimize the traffic on the unprotected waterproofing material. The weight per inch is 0.25 lb/ft^2.

FABRICS

Most roofing manufacturers produce material designed to protect the roofing materials from damage through the installation of overburden materials such as gravel ballast, pavers, and green roofs. These products are typically rated by ounces per square foot or grams per square foot. While this information is readily available from each of the manufacturers for the products, there may not be published data regarding the saturated weight of the product as the water retained by the fabric is likely to be minimal.

Moisture-Retention Materials

FABRICS

Geo-textile fabric moisture-retention products commonly are used in the agriculture and horticulture industries. These materials absorb water when it is available and may store it for plant hydration. There is some difference of opinion among horticulturists as to whether water stored in these products is actually available to the plants when this material is positioned below a root barrier, as plants only uptake water through direct contact with roots. Some of these moisture-retention materials, like rock-wool blankets, hold an impressive amount of water. Depending on the climate, these moisture-retention materials can remain wet for long periods, which could be detrimental to sedums and drought-tolerant succulents as they cannot tolerate wet roots for extended periods. In arid climates, however, a quality moisture-retention pad, positioned above the root barrier, can significantly reduce the frequency of supplemental irrigation.

GEL PACKS AND PARTICLES

These starch-based products are available in a variety of application strategies: packed in packets that are laminated to a geo-textile, encased in permeable pouches that are strategically located near plant root zones, and in particulate form that can be uniformly blended into the growth media. There are a few concerns regarding the use of these products. These products absorb several hundred times their volume in water and discharge the water slowly; however, there is a finite period (typically it is less than 10 years) that starches can cycle through hydration and discharge. While these products may be useful for establishment of the green roof plants, they may not be effective for long-term irrigation strategies. The hydration characteristics of the starch products create other concerns. Since these products absorb many times their volume of water, they expand considerably when hydrated. The bloated material expands and can displace growth media. When the material discharges water and returns to a dehydrated state, voids can be created in the growth media reducing the capacity of the growth media to buffer the effects of temperature extremes. Finally, since these products were developed for extended hydration of potted plants that are manually watered to saturation levels, there is some concern among horticulturist that the rapid hydration of these materials would actually draw water away from the plants during small rainfall events. Many of these products will list their saturation weight in their product literature; some may require some calculations based on volume capacity.

EGGSHELL AND DIMPLED MATS

These products typically serve to provide passageways below the growth media for water to move laterally across the roof surface facilitating drainage from the rooftop. The matrix of cups or dimples may fill with water and act as a reservoir for the plants. However, unless this water is made available to the plant roots by positioning above the root barrier, the plants may not be able to take up the water. Some evidence suggests that as the water evaporates from the reservoirs and passes through the root zone as water vapor, the cooling effect may help retain moisture in the growth media; promoting plant growth without actually hydrating the plants. Research looking at the effect of this material on plant performance will be discussed later. For engineering purposes, one must determine the volume of water retained in each square foot of the material included in the green roof design. When this information is not supplied by the manufacturer, some simple geometrical calculations will produce the weight data needed.

FILTER FABRICS

Filter fabrics prevent particulate from exiting the green roof with storm water run-off and entering the drainage system. Often filter fabrics contain chemicals that repel root growth and serve as root barriers as well to retain growth-media particles. These products are typically lightweight and do not retain a great deal of water. Consult the manufacturer's literature to obtain weight data.

Root Barriers

FABRICS

As mentioned previously, these are filter fabric materials that contain chemicals that repel root growth to keep plant roots from damaging the roofing or waterproofing materials. Fabric root barriers are designed to repel root growth on smaller plants and are best suited to shallower green roofs propagated with Sedums and succulents with fibrous root structures. These materials typically do not add significant weight to the green roof design. Consult the manufacturer's literature to obtain weight data.

THERMAL PLASTIC

Thermal plastic root barriers are more expensive than fabric root barriers but provide protection from root penetration from larger green roof plants like small trees, shrubs, and native grasses that have more aggressive taproots. These materials have similar characteristics to thermal plastic roofing membranes and once the seams are heat welded, form a monolithic surface impervious to water and root penetration. Therefore, these materials must be positioned below the drainage layer, typically just above the waterproofing material.

Drainage Layer Materials

AGGREGATE

Aggregate drainage layers, while popular in Europe, are not widely used in the United States. Conveying and distributing aggregate across the rooftop is labor intensive, and it is more expensive to install than geo-textile drainage products that simply roll out over the roof surface. Aggregates can exceed 4 lb/ft^2

for every inch of depth. Few projects can support this additional weight and, given the added installation cost, aggregate drainage layers may not be an attractive alternate to geo-textile drainage products.

GEO-TEXTILES

As stated previously, geo-textile drainage products are lightweight rolled goods that simply unroll out over the roof surface. There are a variety of configurations available; the most popular is a matrix of cups formed into a plastic sheet. The cups vary in size, height, and spacing among different manufacturers, but all serve to provide passage ways to allow water draining through the growth media to move laterally across the roof surface and enter roof drains and gutter systems. Differing from the cup matrix, there are products that utilize grate and weave configurations that provide drainage passageways, but do not store water. The weight of these materials, as well as the saturated or filled weight of the cup matrix products is available from the manufacturers.

COMBINATION DRAIN CORE/ROOT BARRIERS

These are the simplest to use and most commonly used products for green roof construction. Combination drain core/root barriers often incorporate a soft-felt filter fabric laminated to the bottom surface of the drain core that serves to protect the roofing materials on which it rests. The cup matrix drain core in the center provides for lateral movement of water across the roof surface. A root barrier is laminated to the top of the drain core. In a single step, the product is rolled out over the roof surface to serve as protection board, drainage layer, and root barrier. There are a variety of manufacturers producing similar products. Most include the saturated weight of their product in their literature. Again, if the information is not readily available, it is easy to calculate using simple geometry to determine the volume of each cup and then multiplying by the number of cups in each square foot of the material. (See Fig. 2.1.)

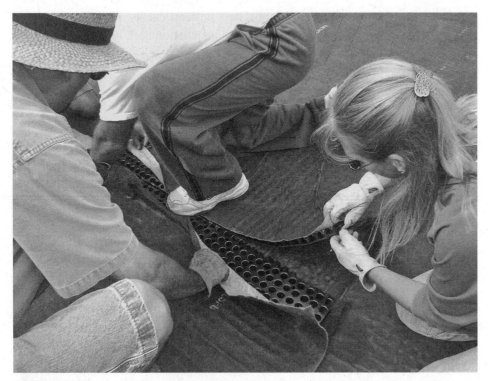

Figure 2.1   Composite drain core combines the soft filter cloth that makes contact with the roofing materials, the matrix of plastic cups to provide passageways for water to move laterally beneath the growth media, and the root barrier on the top, all laminated together making for installation of three products in a single application.

Growth Media

COMMERCIAL BLEND

There are several suppliers of commercially marketed green roof growth media. These formulations can be more expensive than custom blends and may include unwanted or unnecessary ingredients. Specific attributes and drawbacks of using these formulations will be discussed later. One benefit, however, is that these suppliers have conducted saturation testing to determine the saturated weight of their growth media formulations and they make this data readily available for consideration.

CUSTOM BLEND

Most green roof growth media consist of blended, expanded aggregate, and organic material. The availability of these ingredients varies regionally. Choosing to custom-blend our growth media leads to greater control over the ingredients and the ability to choose regionally produced materials over those that must be transported long distances. There will be saturated weight data available for manufactured aggregates, but for natural aggregates like pumice and volcanic rock it may be necessary to conduct saturation tests. Sustainable growth media formulations are typically blended in ratios of 80% aggregate with 20% organic materials. One can get a reasonably accurate saturated weight of the custom-blended growth media by substituting the saturated weight of the aggregate for the organic portion of the blend. One can also conduct saturated weight-testing of a custom blend using the following steps: Submerge a (12 inch × 12 inch) sample at the desired depth for 24 hours, remove the sample and allow to drain for 15 minutes, weigh the sample, and subtract the tare weight of the container to determine the saturated weight of 1 square foot of the custom blend at the given depth.

Plant Material

Plant selection for the purpose of structural loading calculations can be broken into the following basic categories and corresponding weights*:

Sedums and succulents—2 lb/ft^2 (see Fig. 2.2)

Grasses and bushes up to 6 inches—3 lb/ft^2 (see Fig. 2.3)

Shrubs and bushes up to 3 feet— 4 lb/ft^2 (see Fig. 2.4)

*International Green Roof Association Global Networking for Green Roofs

Figure 2.2 Simple planting strategy made up of sedums and succulents thrive in growth media depths ranging from 4 to 6 inches and require minimal maintenance.

Figure 2.3 Grasses and bushes up to 6 inches in diameter thrive in growth media depths ranging from 6 to 8 inches. It is possible to support this plant group in shallower growth media depths by supplementing hydration requirements with artificial irrigation.

Figure 2.4 Shrubs and bushes may require growth media depth of 12 inches or more and, depending on the climate, may require supplemental irrigation.

Drawings

It is now time to put the design into a format that conveys the vision to members of the green roof design and construction team. Even if the team is made up of only one person, it is important to organize the design and concepts using drawings. Always retain each draft so that you can look back at the evolution of the project. Necessary changes will be easier if you can identify the phase you need to revisit. Some designers even keep several sets of drawings depicting the different options and the subsequent direction each decision guided the process. Sketching the cross-section will identify each of the components and their position in the green roof design. One can get a good idea of the weight and cost of the green roof design through the development of the cross-section. A sketch of the roof-plan view, or bird's eye view, of the green roof is also necessary. Through the roof plan the plant layout is developed, the various features are identified, and the construction process is organized. (See Figs. 2.5 through 2.9)

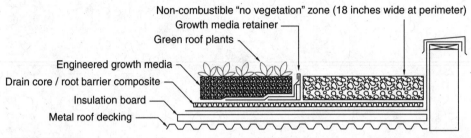

Figure 2.5   **Typical cross section built-in-place green roof.**

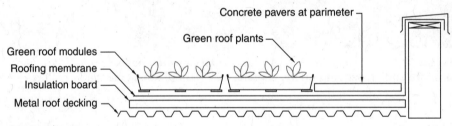

Figure 2.6   **Typical cross section modular green roof.**

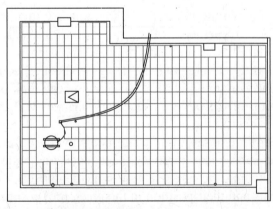

Figure 2.7   **Roof plan simple modular roof.**

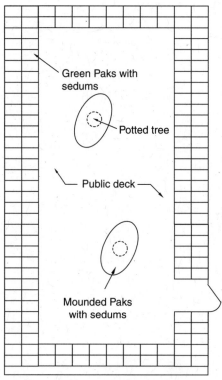

Figure 2.8 Roof plan typical modular roof garden.

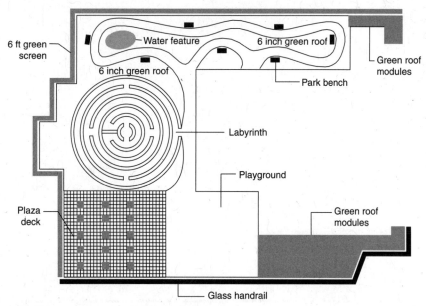

Figure 2.9 Roof plan multi-use rooftop garden.

3

Waterproofing

The roofing industry uses the term *waterproofing* to refer to liquid-applied sealants used to weatherproof concrete substrates. The term *roofing membrane* refers to roll goods that are typically used to weatherproof various types of roof substrates. To simplify discussion, the term *roofing* is to encompass all types of material used to weatherproof the roof substrate for a green roof. There are almost as many roofing options for green roofs as there are plants. Certain project conditions make some roofing strategies better suited for green roof applications than others. This section will discuss the characteristics of various roofing options to help identify the strategy best suited for a green roof project.

Liquid-Applied Membranes versus Roll Goods

Liquid-applied materials like hot modified rubber and elastomeric urethanes cure to form a monolithic membrane with no seams. Some systems are multilayered using interplay fabrics or felts to increase durability. These systems are widely used to weatherproof concrete substrates. The material is applied directly to the concrete to form a membrane that is fully adhered to the roof deck. Fully adhered membranes eliminate lateral movement of water between the membrane and the roof deck. If the membrane is damaged, then there is no place for the water to go except through the concrete. This makes the damaged area easier to locate, based on the location of the leakage on the interior of the building, and reduces the interior water damage. The liquid material has a maximum bridging capacity, which can be problematic for larger fissures and cracks in the concrete substrate. It is critical to plan for building expansion to reduce the risk of cracks and tears at points of building movement. Roof penetrations are always potential problem areas. When possible, penetrations should be kept to a minimum. When penetrations are unavoidable, flashings must be installed with great care and thoroughly inspected prior to proceeding with the green roof assembly. The rapid installation of liquid-applied roofing systems affords large projects economy of scale, making this roofing strategy less expensive than rolled goods for these large, concrete structures.

The most common roofing materials are produced in rolls of various lengths and widths. These materials are installed by adhering, mechanically attaching, or loose-laying and ballasting in place over the roof surface. Each of these installation strategies has attributes and drawbacks. Fully adhered membranes are typically more expensive due to the cost of labor and material for the adhering process. Fully adhered membranes minimize the migration of water below the membrane in the case of roof leaks; thereby minimizing damage to roof insulation and the building interior. Pinpointing the source of leaks is usually easier on fully adhered roofing systems because leaks typically show up on the interior of the building near the area of the roof damage. Mechanically attached roofing systems were developed as a cost-saving alternative to fully adhered roofing systems. These roofing systems are typically fastened along the edge of the sheet using screws with large sheet metal washers called *plates*. The screws and plates are concealed by overlapping the adjacent sheet over the fastened edge covering the screws and plates. This is a very cost-effective installation strategy because the rolls of membrane can be positioned and fastened rapidly at significant labor savings. Though the lower cost makes mechanically fastened roofing systems an attractive choice for a project, this strategy has an inherent characteristic that makes this installation option unsuitable for green roof projects: the screws. The screw heads lie just below a single layer of the roofing membrane. The weight of the green roof system presses down on the membrane, exerting pressure where the membrane makes contact with the screw head and the plate. Building vibration and harmonic movement can prematurely wear on the membrane that is pressing against the screw and plate. Additionally, some compression of the insulation can be expected, causing the screw to sit higher than the insulation, poking the underside of the membrane. This can also be a problem for fully adhered roofing installations and is remedied by either covering any mechanically fastened insulation with a cover board set in adhesive or by setting the roof insulation in adhesive instead of using mechanical fasteners. The last roofing configuration is loose-laid and ballasted membrane, where the roof assembly is simply laid over the roof surface and then weighted in place by ballast, in this case, the green roof system. This is the most cost effective means of roof installation, eliminating the labor and materials associated with fully adhering or mechanically attached membranes. One drawback to loose-laid and ballasted roof installation is lateral migration of water in the case of a roof leak, making the source of the leak difficult to trace and increasing the amount of water damage to the roof insulation and building interior. Another drawback is that ballast is required over the entire roof area to secure the roofing system. This can make this installation strategy unsuitable for projects wishing to utilize modular green roof systems to complete the green roof in phases; as the roof area not greened would require temporary ballast, which would then require removal before future phases could be greened.

Overkill and Prudent Design

Green roofs are new in the United States, and thus many of the people involved in their design and construction are operating outside of their comfort zones. The resulting impulse usually is to attempt to design for every contingency, and in doing so people often cross the line of prudent design and tumble into the land of overkill. This is a huge problem that is plaguing the green roof industry. If everyone who touches the project throws in extra money or premium materials, the end result is a green roof that comes in over budget and never gets built. Let's take a closer look at where these overkill decisions occur.

The green roof designer decides which materials to use and from which manufacturers they will be purchased. The roofing industry relies heavily upon these manufacturers to supply long-term warranties on the roofing systems purchased. The industry has developed the term *no dollar limit* (NDL) warranty to refer to a warranty that has no monetary limit on repairs covered under the warranty. This is a relatively new development in the roofing industry over the last 15 to 20 years; it began about the same time single-ply roofing systems entered the market. Prior to NDL warranties, building owners relied on third party underwriters to issue coverage; for example, you may have heard of a 20-year *bonded* roof. These were installations that were underwritten by surety companies. The term *bonded* is still widely used in contractor advertisements, although the term now refers to the company's ability to secure a performance bond for a particular project rather than blanket coverage. During the transition period prior to the NDL era, many building owners engaged the services of an independent roofing consultant to oversee the installation process and monitor quality control measures. The NDL warranty provides coverage for both workmanship and material deficiencies, suggesting that the role of quality control is now assumed by the manufacturer. This means a much larger risk for the manufacturers issuing the warranty. Additionally, there are government mandates that require manufacturers to place funding for warranty exposures in reserves, tying up precious capital. Therefore, manufacturers strive to sell premium materials at higher profit margins with less risk of warranty claims.

When looking to the manufacturer to specify a roofing system under a green roof that they will have to warranty, they have little regard for bringing our project in on budget. If this just meant spending extra money on a better membrane, it might be considered a worthy investment. However, overkill rarely stops at the membrane. Most manufacturers insist on using premium assemblies when specifying roofing systems for green roofs. Typically, this includes premium materials including insulations, adhesives, vapor retardants, flashings, and edge treatments. Many manufacturers have developed multilayer redundancy within the roofing systems they condone for usage under a green roof. Some manufacturers have developed proprietary drainage and root-barrier products that are required to receive NDL warranty coverage. The end result is a high-end roofing system with a big price tag.

Some designers may feel that as long as the roofing manufacturer is issuing a warranty that eliminates the financial risk for the building owner, the security is worth any price. However, reading the fine print reveals that most membrane manufacturers insert exclusion clauses in the NDL warranty for expenses related to the removal and reinstallation of the growth media and green roof plants to conduct warranty service work. This is a significant expense, and with the exception of complete membrane failure, it is typically more expensive than the repair work itself. This clause is commonly referred to as the "removal of overburden" clause. The overburden clause renders the NDL warranty less-than-effective in eliminating the financial risk of material failure when dealing with green roof applications, leaving the building owner with an overly expensive roofing system and less-than-complete warranty coverage.

Just as single-ply roofing systems prompted a change in the way the roofing industry underwrites warranties, perhaps green roofs require green roof designers to rethink the value of the NDL warranty for green roof projects. Perhaps green roofs do not fit into the typical paradigm; this is where one must use common sense and choose membranes that are familiar and that have performed well on non-green roof projects. If the green roof designer specifies quality roofing systems, employs reasonable quality assurance measures, awards the work to reputable contractors, and exercises due diligence in membrane testing prior to installation of the green roof components, a successful and leak-free green roof project can be constructed at a reasonable price.

Unfortunately, the roofing industry is not the only body of suppliers and contractors bumping up their fee structures to compensate for their unfamiliarity with green roofs. Engineers, growth media suppliers, nurseries and greenhouses, landscape contractors, and even hoisting operators are all adding money to their normal fee structure when they know that the project is a green roof. The market will correct most of this as the number of green roof projects continues to increase. Until then, designers have to compare the prices of similar goods and services not related to green roof construction with the bids they are holding in their hands. If the numbers are not in line, then the supplier or contractor needs to know. Many times it may be just a matter of discussing the project in-depth to help eliminate some of the uncertainty. To spot some of the areas where the price has been inflated, check for delivery cost and truck standing charges, disproportionate labor figures, higher-than-necessary planting densities or plant sizes, and overly aggressive maintenance regimes. Every time somebody boosts the price due to inexperience, it makes the task of getting the green roof built that much more difficult.

Sunshine

One of the primary returns on investment for building a green roof is the resulting longer life of the roofing system. Extended roof life is listed in practically every news story, magazine article, and advertisement that praises the benefits of green roofs. It is true that the two most destructive forces on any roofing material are the ultraviolet sun rays that degrade the material's ability to expand and contract with temperature changes and the dramatic daily temperature fluctuations that exploit the material's deteriorating flexibility. Green roofs obviously block exposure to the sun's harmful rays and, as many studies have shown, eliminate the heating of the roofing material during the day and the rapid cooling at night and keep the material temperatures relatively constant, even keeping the membrane warm during winter months. Absent the UV exposure and effects of freeze-thaw cycling and extreme daily temperature fluctuations, the life of the roofing material can be extended to three or four times that of exposed material. However, this life-span benefit only applies to the roofing material protected by the green roof. If the project is only installing enough green roof area to cover 30% of the roof to meet a governmental mandate, for instance, the area of the roof protected by the green roof will enjoy extended life while the other 70% will fail within the typical life-cycle limitation of the material. This is often seen with modular green roof strategies because the self-contained nature of the system allows modules to rest on the roof independently without requiring additional retainers or containment methods. Some projects even have a phasing plan to gradually cover the roof area over time. With some strategic planning, one can use sacrificial plies of roof membrane, roof coatings, gravel, or pavers to help protect these exposed areas extending the life cycle of the roofing material. Many of the projects that cover 100% of the roof area with the green roof neglect to protect the vertical surfaces of the curb and wall flashings. Gravity pulls at vertical seams and makes these areas potential problem spots for all roofs and warrant special attention for green roof installation. Doubling the flashing material allows the exposed layer to weather, keeping the base layer in pristine condition. Routinely coating these materials with reflective roof coating will add additional life and will help to ensure a leak-free green roof for the extended life cycle so widely touted in the media.

Water Testing

Many green roof specifications call for flood testing of the roof prior to installing the green roof components. This procedure can be troublesome and stressful for engineers and roofers alike. Sometimes the slope of the roof is such that to have 2 inches of water standing on the high point of the roof, the water is frighteningly deep at the low point. Roofers worry about the fact that no roof construction is designed to withstand water entry pressurized by the immense weight of so much water, while the engineer worries about the immense

Figure 3.1   Flood testing of the roofing system prior to installation of the green roof components ensures the roofing is leak-free.

weight of so much water. There are no easy answers here; the flood test can give everyone a sense of confidence that the roof has no leaks and is ready to receive the green roof. It also helps to relieve the roofing contractor of leak responsibility by having demonstrated the installation to be leak-free at the commencement of the green roof installation. (See Fig. 3.1.)

Some large projects have used sandbags to section off portions of the roof to conduct flood testing in smaller sections, while others have used fire hoses to spray water systematically over the roof. There are also more sophisticated leak-detection methods such as nuclear, infrared, that identify moisture present beneath the roofing materials and electronic vector-testing that uses electric current to find breaches in the surface of the roofing materials. Regardless of the testing method, it becomes a moot point if the roofing is allowed to be damaged during the construction of the green roof. It is critical to discuss protection methods and observation responsibilities with the green roof construction team during the preconstruction meetings. Each employee working on the roof surface must be educated about protection procedures and damage notification procedures. It is all too common for a new employee to accidentally damage the roofing and be hesitant to report the damage. Often the damage is concealed, only to show up on the interior of the building some months after the green roof is completed. Each employee must be taught that the real mistake is not to have the accident that damages the roofing, but to allow the damage to go unreported. Repair of a minor puncture costs typically less than $100; however, that same repair can cost thousands after the green roof already has been installed.

4

The Dirt on

Green Roof

Soil

The most critical component of a successful green roof and the most common cause for green roof failure is the green roof soil or growth media. Some early green roofs were constructed by professionals with garden and landscape expertise using soils and planting strategies that brought them success in the construction of planter boxes and landscaping projects at ground level. Relying on these soil formulations and planting strategies produced some highly publicized green roof failures. This chapter discusses successful growth media formulation and will identify some of the common pitfalls.

Quantity and Composition

During the design phase the identified plant group chosen for the project determines the depth of the growth media.

To determine the quantity of growth media required for the green roof project, one must first calculate the area of the green roof in square feet. Next, use the depth of the growth media in inches to calculate the volume of total growth media required for the project. The volume calculation is easy when the growth media depth is an even factor of 12:

4-inch depth: (square feet of area/3 = cubic feet of growth media)

6-inch depth: (square feet of area/2 = cubic feet of growth media)

9-inch depth: (square feet of area/1.33 = cubic feet of growth media)

If the growth media depth is not an even factor of 12, then one can find the volume by first converting the square feet into square inches by multiplying the area in square feet by the depth in inches and then by 144. Next, divide by the number of square inches in a cubic foot, which is 1728 cubic inches.

$$(\text{Square feet of area}) \times (\text{depth in inches}) \times 144)/1728 = \text{cubic feet of growth media}$$

Once you have the volume in cubic feet of growth media required for the green roof, you will need to convert to a unit of measurement commonly used by the industries providing the ingredients of growth media: cubic yards. To convert cubic feet to cubic yards, one simply divides by 27—the number of cubic feet in a cubic yard.

Example: Calculate total volume of growth media.

Depth = 7 inches; green roof area = 100 feet × 50 feet = 5000 ft^2

Growth media (in cubic feet) = $(5000 \times 7 \times 144)/1728$ in^2/ft^3

$$= 5{,}040{,}000 \text{ in}^2/1728 \text{ in}^2/\text{ft}^3$$

$$= 2916 \text{ ft}^3$$

Growth media (in cubic yards) = 2916 ft^3/27 ft^3/yd^3

$$= 108 \text{ yd}^3$$

Now that the total volume of blended growth media required for the green roof project has been calculated, it is time to identify the source of the material. If a commercially marketed, ready-mixed growth media is being used, orders based on the calculated quantity can be placed. There are several suppliers to choose from, so there is an opportunity for comparison shopping at this stage. The makeup of these proprietary growth media formulations vary somewhat and may not be particularly well-suited for a specific green roof project. Many of these formulations include ingredients like sand, peat, humus, and organic compounds that may not be desirable for the growth media. Most of these producers claim that their growth media is blended to the standards established by the authoritative body on green roof construction in Germany, referred to as FLL. Without discussing the accuracy of these claims, the point may be moot as green roof design factors are radically different in Germany than in the United States.

What is needed is a lightweight, sustainable growth media that will support plants for many years while not overloading the structure. In order to achieve long term sustainability, the growth media must contain high percentages of mineral material that will not break down over time. Expanded aggregates, pumice, and volcanic rock are lightweight aggregates with pore spaces capable of holding the water necessary to support the plants. However, plants require some organic material for nutrition. The ratio of mineral to organic material for successful green roof growth media is 80% (or more) mineral to 20% (or less) organic. Organic material will breakdown in a relatively short period of time. Higher percentages of organics will result in a loss of depth as the organic material breaks down. The 80/20 ratio provides enough organic material to establish the plants while maintaining the desired depth for the life of the green roof. The organic material in the original growth media blend will decompose in 3 to 5 years. However, the foliage that sheds from the plants will lay on the surface of the growth media, decomposing to continually recharge the organic requirement of the growth media and maintaining the required cationic exchange for the plants to thrive.

Sourcing Ingredients

Sourcing the components of the growth media on your own and having the mixture custom blended can result in significant savings. Often the materials can be sourced locally, helping to achieve the LEED credit for regionally manufactured materials. Start by identifying the aggregate for the media and finding the source. There are three varieties of expanded aggregate: expanded clay, expanded slate, and expanded shale. While these materials are manufactured products, there are also naturally occurring aggregates like pumice, scoria, and volcanic rock. These materials are largely interchangeable. Ongoing research directly comparing the performance of various aggregates is discussed in Chap. 11. Each of these materials is not, however, readily available in all regions of the country. There are areas where expanded shale is the dominant aggregate and none of the natural aggregates may be available. Transporting aggregates outside of their region adds to the cost of the aggregate and often raises the final cost higher than other aggregates that are readily available within that region. Therefore, the first task is to find the aggregate regionally available with the lowest cost. Most producers have several gradations to choose from. The desired gradation for most green roof projects will have particle sizes ranging from 1/8 to 1/2 inch with minimal fines.

Finding a Blender

Next one has to find the organic materials and someone to blend the mixture; this is typically the same company. Often suppliers of compost, mulch, and soil blends are equipped with the necessary means of receiving a shipment of aggregate, supplying the organic material, blending the mixture to one's specifications, packaging the blended growth media for the desired mode of transportation, and delivering the finished product to the project site. Many of these businesses are enthusiastic about providing their goods and services to the emerging green roof industry. Again, there are a variety of organic materials available for use in growth media, and these materials vary in availability from region to region and from supplier to supplier. Some of the higher-quality materials include composted pine bark, worm castings, and composted coconut core, but many times composted yard waste may be the only material available. There have been some issues with projects using composted yard waste that contained residual chemicals from pesticides and herbicides, so it is important to inquire about the supplier's quality control measures when yard waste is the only available material. Reputable suppliers exercise due diligence in providing quality compost that has been processed for the proper durations necessary to eliminate chemicals that can be harmful to green roof plants such as testing compost per Test Methods for Evaluation of Composting and Compost (TMECC) as identified by the United States Composting Council. The contamination of the compost with weed seed that then germinates in the green roof and causes maintenance problems can be another issue. Most suppliers have stringent protection protocol, such as keeping stockpiles covered or stored indoors, to keep weed seed out of the compost and subsequently out of the blended growth media. It is important to continue this protocol through the delivery and jobsite storage phases.

Transporting and Hoisting Blended Growth Media

Growth media is heavy and bulky material and usually requires large commercial vehicles for transport to the jobsite. The type and size of vehicle required is typically determined based upon the volume of material required for the project. The growth media for small- to medium-sized green roof projects can be placed in 50-lb bags or large super sacs which hold 2 to 3 yd^3. Either of these can be palletized for fork-lift handling and can be transported to the project using trucks ranging from pick-ups to tractor trailers. Larger projects can eliminate the cost of bagging and palletizing by using dump trucks to deliver up to 44 yd^3 of growth media per load. (See Fig. 4.1.)

Figure 4.1   Ordering growth media in bulk can reduce the cost.

Once the material arrives at the jobsite, it must be conveyed to the rooftop. Many projects incur unnecessary expenses due to improper coordination of the delivery and roof-loading of the growth media. It is critical to understand the setup procedures and durations of various hoisting and conveyance equipment. When scheduling manpower, it is important to stagger the starting time for the workers to eliminate paying for wages for time spent waiting. It is equally important to schedule deliveries with conveyance rates throughout the day to eliminate paying for both unproductive wages and standing charges from trucking firms. Though cranes may be necessary for material conveyance to rooftops greater than 40 feet above the ground, the roofing industry has alternative strategies that can significantly reduce project cost for projects that are less than 40 feet above the ground. For example, many roofing contractors have rooftop hoisting equipment and conveyors that they use to convey gravel ballast to the rooftop. (See Fig. 4.2.)

Figure 4.2 Material-handling equipment and techniques used by the roofing industry to convey conventional roofing materials to the rooftop can be employed for green roofing materials as well.

The efficiency with which the roofing industry conducts material handling for rooftop construction is second to none. Including the roofing contractor in this phase of the project can produce cost-saving synergies between trades and reduce duplication of efforts. For example, coordinating the unloading of the roofing equipment with hoisting of green roof components can eliminate one of the days of hoisting charges. Common use of fall protection, labor forces, setups, and breakdowns are other opportunities for cost savings.

Another popular conveyance method for green roof growth media is the blower truck. This large-scale vacuum equipment draws growth media through a flexible hose to the rooftop, where a worker can spray the material in the desired depth across the green roof area. The use of this equipment, although quite expensive, dramatically reduces the labor necessary to distribute growth media. There is a process called *hydroseeding* that combines growth media and plant seed and employs blower technology to distribute growth media and propagate green roof plants in a single step. This strategy has the potential to produce large expanses of green roof at a fraction of the cost of manual methods. However, the more advanced equipment required for this process is slow in making its way from Europe into the U.S. market place. As U.S. demand for green roofs increases, competing producers and importers of this equipment will reduce the cost and increase their numbers, helping lower the cost of the hydroseeding process.

5

Construction

There are many commercial green roof systems and private designers, and green roof construction methods vary widely among providers. The purpose of this chapter is to consider several strategies for green roof construction and identify conditions that make certain strategies better suited for a specific project. Ultimately, the goal is to design and build a green roof that meets the project's intent, is completed on time, and stays within budget. A guiding principle that will help attain this goal is to "keep it simple." The success of a project lies somewhere along the delicate balance between what is adequate and what is overkill.

Modular or Built-in-Place

Modular green roofs are green roof systems composed of planters that are arranged on the rooftop. Most of these products

Figure 5.1 Modular green roofs are green roof systems composed of planters that are arranged on the rooftop.

include engineered soil blends and plants based on the regional climate of the project. This method makes the design and installation of the green roof very simple as the modules are self-contained and arrive to the project ready to be set in position on the rooftop. (See Fig. 5.1.) There are several manufacturers who market green roof modules constructed of various materials and in various configurations. Preplanted modules can be conveyed and positioned on the rooftop, allowing for the completion of the green roof with a single application. This makes modules particularly attractive for rooftops with limited or difficult access. Modules can be moved easily, allowing for ready access to the roof surface. This feature helps to alleviate concerns over the repair of roof leaks or future roof alterations. The free-standing nature of the modules allows the edges of the green roof to be defined without the use of edging or barriers to contain the growth media. This means that the green roof can be phased: adding modules over time or as budgets allow.

Modules constructed of rigid material are more difficult to fit into irregularly shaped roof areas, while fabric modules provide greater flexibility for these projects. However, rigid modules can be grown to maturity offsite and delivered to the project to produce an instant green roof, whereas fabric modules must be grown in place on the rooftop. (See Fig. 5.2). The growth

Figure 5.2   Green roof modules constructed of rigid material are more difficult to fit into irregularly shaped roof areas, while fabric modules provide greater flexibility for these projects.

media remains sealed within the fabric module until conditions are optimal for planting, allowing for construction of the green roof during any time of the year without the use of erosion blankets and wind screens. (See Fig. 5.3.)

Most modular systems are extensive green roofs with growth media depths less than 6 inches and plant palettes consisting of low-lying ground covers and succulents. Some modules are available in deeper units; however, the weight of deeper modules often renders them immobile once filled with growth media. These deeper modules are capable of supporting a wider variety of plants, including native grasses, bushes, and small trees. A blend of shallow extensive modules with deeper modules strategically positioned provides some definition to the green roof with the taller plants and produces a more natural green roof setting. Concrete and synthetic paver systems can be used in combination with green roof modules to create

Figure 5.3   Rigid green roof modules can be grown to maturity offsite and delivered to the project to produce an instant green roof.

paths and walkways through the green roof as well as to create sitting and gathering areas. (See Fig. 5.4.)

The self-contained nature of green roof modules makes them particularly well suited for green roof research projects. Modules can be easily outfitted with different growth media and planting strategies to produce replicated experiments. These individual modules can then be monitored under desired conditions and factors to help us learn more about green roofs. The affordability of the modules also often makes this green roof strategy attractive for small-class projects and school green roof displays. (See Fig. 5.5.)

Although green roof modules may offer a simplistic approach to green roof construction, they may not be the best fit for every project. Intensive roof gardens, for example, require much deeper soil to support a more diverse plant palette that includes native plant species, bushes, and small trees. (See Fig. 5.6.) These rooftop gardens are landscaped using techniques and features typical of landscaping projects

Figure 5.4   Concrete and synthetic paver systems can be used in combination with green roof modules to create paths and walkways through the green roof as well as to create sitting and gathering areas.

Figure 5.5   Students at Southern Illinois University Edwardsville conducting green roof research. Green roof modules lend themselves to research projects that require replicates of each variation under study as each module represents one replicate.

Figure 5.6   Roof gardens that employ grasses and larger plants may be difficult to construct using green roof modules and typically are built in place by layering green roof components and growth media to form uninterrupted planting surfaces.

conducted at ground level and can include water features, stone accents, labyrinths, playground equipment, among other features. (See Figs. 5.7 and 5.8.) These rooftop gardens are usually designed for use over structural concrete roof construction with very high loading capacities. Some underground structures are designed for seamless transition with the surrounding landscaping, completely concealing the building section that lies below the green roof. These natural settings may be difficult to accomplish using the modular approach and may require the more traditional built-in-place construction where the green roof assembly is constructed by layering materials in place over the roof surface. The next several sections will discuss materials used in built-in-place green roof construction.

Figure 5.7   This basketball court is situated over a grocery store in Germany, providing public access to recreational space otherwise not available at ground level in this community.

Figure 5.8   In this community with limited space, the parking garage is situated over two levels of commercial space containing a grocery store and office space, and the green roof containing a playground and basketball court is on the rooftop.

Roof Slope

While many green roof projects appear to be constructed on flat rooftop surfaces, the fact is that almost all roof surfaces must have some degree of slope to facilitate proper drainage. Some exceptions, however, are older structures that were actually built with rooftops that are dead level. The roofing systems on these older flat roofs were made from coal tar pitch, a material with a very low melting point that liquefies each year in the summer heat to self-heal cracks and splits that occur during the winter months. The use of coal tar pitch has largely been replaced by asphalt bitumen roofing systems and modern single-ply roofing membranes. Many of the older coal tar pitch roofs have been retrofitted with tapered insulation to provide the necessary slope for modern roofing systems. For most roofing manufacturers the minimum slope is 1/8 in/ft, although most designers have moved to 1/4 in/ft roof slope to accommodate some deflection of the roof structure. Proper roof slope is critical in facilitating the complete drainage of water from the roof surface because water that is allowed to pond on the roof dramatically accelerates the aging process of the roofing material. In fact, many roofing manufacturers' warranties have exclusions written into the language for material failures in areas of the roof that allow water to pond.

Rooftops with slopes that exceed 2 in/ft are considered to be steep roofs. Steep roof slopes are often designated by terms indicating the rise of the roof slope over the horizontal distance covered by the roof. For instance, a roof surface that rises 4 inches for each 12 inches of horizontal building area would be termed 4/12; this designation sometimes is referred to as rise over run. (See Fig. 5.9.) Green roof projects for steep roofs (slopes exceeding 2/12) require additional design consideration to keep the growth media from sliding down the roof slope. While there are a variety of manufactured products for this purpose, there are also some bracing strategies using treated lumber that can be designed into the project. These strategies all compartmentalize the growth media into honeycomb or grid structures that are anchored in place. The growth media fills each compartment and supports the plant life that will grow to conceal the anchorage structure. (See Fig. 5.10.)

Figure 5.9 Green roofs are not limited to flat roofs evidenced by this Nordic style roof.

Figure 5.10 Bracing strategies such as grids constructed of treated lumber or manufactured plastic products molded into honeycomb configuration help hold green roof growth media in place on green roof projects on steep-sloped rooftops.

Figure 5.11   Prevegetated green roof modules can also be utilized to create green roofs over steep-sloped rooftops. This project uses the stainless steel clamps to anchor the handles of the modules together, helping to keep the modules from sliding down the roof slope.

Modular green roofs often can be used on slopes as great as 4/12 without additional bracing. Using modules on steeper slopes may require additional design considerations like connecting modules together and to a central anchor point along the top of the roof slope. Additionally, the modules may be grown to maturity offsite to allow the plant roots to bind the growth media for added stability. (See Fig. 5.11.)

The Location of the Insulation

A major component of roof construction is the roof insulation. There are three basic strategies for roof insulation that are differentiated by the location of the insulation. Most homes are insulated with soft insulating material, like fiberglass or cellulose, placed just above the ceiling. The roof is then constructed with a ventilated attic space between the insulation and the roofing. This method, commonly referred to as cold roof assembly, allows for an equilibrium between the temperatures above and below the roof, reducing condensation and ice damming.

Warm roof assemblies locate the roof insulation either just above or just below the roofing material; this method is most commonly used in flat roof construction. Many green roofs are constructed using rigid insulation boards made of polyisocyanurate (also known as *iso*). This insulation must be kept dry and therefore must be installed below the roofing material. This methodology, the most common for flat roofs in the United States, leaves the roofing material exposed to the sun and weather elements, unless of course the membrane is protected by a green roof.

Roof insulation made of extruded polystyrene (such as the pink- or blue-colored insulation manufactured by Dow Chemical Company) is impervious to water penetration and can be placed on top of the roofing material. This method requires a form of ballast-like concrete pavers or a green roof to hold the insulation in place. This strategy has some benefits over traditional under-the-roofing installation. The insulation can be extended vertically to insulate the curbs and walls and covered with sheet metal or additional roofing material, keeping the roofing material insulated from UV exposure and wide temperature fluctuations, which will extend the life of the vertical flashing material. Dow Board has a higher density rating than iso; that is, it has higher compression strength than iso insulation. This can be an important characteristic when rolling heavy equipment across the roof surface. Iso insulation has a tendency to crush or collapse under the weight of such traffic, and iso is especially vulnerable to damage from repeated travel over the same pathway. Care must be taken to protect the roof insulation with plywood to help distribute the weight over a larger area of the roof surface. A more subtle benefit of this roofing strategy, however, is the ballast requirement. A green roof that is designed into the construction of the building as an integral component, such as the roof ballast, is less likely to be eliminated from the project through value engineering should the project exceed its budget. If utilized as ballast, elimination of the green roof would require a complete redesign of the roofing system. The resulting additional architectural and engineering fees would negate some of the savings, making the loss of the green roof as an amenity less attractive.

Drainage

Many European green roofs are constructed using a layer of granular material, like gravel, under a layer of filter fabric that keeps fine particulate from migrating from the growth media. Water percolates through the growth media, passes through the filter fabric, and moves laterally across the roof surface through the gravel. This porous layer beneath the growth media is essential to allow excess water to drain away from the green roof. Many of the succulents used to propagate green roofs require well-drained soil to prevent disease. It is equally important to begin with a roof surface properly sloped to adequate drainage devices. Ponding water can both promote disease and prematurely age roofing materials.

Granular drainage layers, while effective for roof drainage, are heavy and tend to demand labor-intensive installation. Geotextile drain core products are much lighter than granular drainage materials and may be better suited to projects with limited structural capacity. These products are manufactured in rolls that are easily conveyed to the rooftop and require minimal labor to install. (See Fig. 5.12.) Many of these products incorporate soft fabrics on the underside, to protect the roofing material, and filter fabric impregnated with root inhibitor on the topside. These drain core/root barrier composite products serve to protect the roofing, provide drainage under the growth media, and prevent damage from root penetration; all in a single application. (See Fig. 5.13.)

There are green roof projects with specific elements that may require individual products to perform the functions of protection, drainage, and root barrier. For instance, intensive green roofs with deeper-rooting plants may require thermal plastic root barriers with heat-welded seams to protect against penetration of more aggressive roots. Large roofs sloping to one edge and draining to a gutter may require drainage products with oversized passageways providing for drainage of the greater volumes of water.

Figure 5.12 Geo-textile drainage products are manufactured into rolls that are easily unrolled into position over the roof surface and are less labor-intensive to install and are lighter in weight than aggregate-based drainage layers.

Figure 5.13 These composite drain core/root barrier products combine a soft filter fabric that protects the surface of the roofing material, matrix of plastic cups or strands that provide passageways to facilitate lateral drainage of water beneath the growth media, and a filter fabric impregnated with root inhibitor to provide a barrier that repels root penetration into the roofing or waterproofing materials.

Detailing Accessories

The perimeter of the green roof must be detailed in a manner that is aesthetically pleasing while still performing its desired function. Green roofs that extend to roof edges without parapet walls must be detailed to retain growth media from leaving the rooftop. While lumber may be used to build up the edge to a height greater than the growth media depth, the lumber often warps and deteriorates and can become a maintenance burden. Products like the Green Roof Edge (www.greenroofedge.com) provide an architectural finish to the roof edge in prefinished sheet metal available in a wide range of colors. This product is available in perforated lengths for green roofs that slope to the edge and drain to a gutter. An integrated filter fabric keeps fine particulate from leaving the green roof while the perforations in the metal allow water to drain from the green roof growth media. (See Fig. 5.14.)

Figure 5.14   Green roofs that drain to an edge with a gutter system require a means of allowing water to drain while keeping the green roof growth media in place. Perforated products used in conjunction with filter fabric can keep fine particulate from entering the guttering and plumbing systems.

Green roofs that stop short of the roof edges must also be detailed to retain the growth media. Many green roofs have gravel vegetation-free zones at the perimeter of the roof and around roof penetrations. Many of these are detailed with commercial products like the Growth Media Retainer (www. greenpaks.com/growthmediaretainer.html) that separate the growth media from the vegetation-free zone. These products are perforated to allow water to flow freely and use filter fabric to keep fine particulate from migrating from the growth media. The Growth Media Retainer employs a strip of noncompressible mesh to hold the filter fabric away from the perforations in the aluminum retainer to maintain a constant flow rate over the life of the green roof. When filter fabric is allowed to press against the perforations in the retainer, fine particulate accumulates at each hole and continually reduces the flow through the material over time. (See Figs. 5.15 and 5.16.)

Figure 5.15   Metal edging materials help green roof designers separate the vegetated area of the green roof from nonvegetated areas of the roof.

Figure 5.16   Some of these products are perforated to allow water to flow freely and use filter fabric to keep fine particulate from migrating from the growth media.

Roof drains are another area that requires detailing to keep growth media from migrating from the green roof and, in this case, from entering the building's plumbing system. Products like the Green Roof Drain (www.greenroofdrain.com) roof drain access chamber allows water to flow into the roof drain and employ integrated filter fabric to keep growth media and plant material from entering the roof drain and damaging the plumbing. This product also employs a strip of noncompressible mesh to hold the filter fabric away from the perforations in the aluminum retainer to maintain a constant flow rate over the life of the green roof. The removable lid allows access to the roof drain for routine maintenance. (See Fig. 5.17.) The success of the green roof depends on the correct application and performance of these green roof accessories. Commercial products must be manufactured of quality materials capable of life spans equal to that of the designed green roof. Contractor-fabricated

Figure 5.17 Roof drain access chamber allows water to flow into the roof drain. Some employ integrated filter fabric to keep growth media and plant material from entering the roof drain and damaging the plumbing.

components must meet the same high standards. Growth media and plant material must not be allowed to enter the plumbing system. At the same time, the flow of water draining from the green roof system must not be impeded. Clogged drains and flooded rooftops can easily overshadow the hard work that produced the green roof. (See Fig. 5.18.)

Preparing for the Plants

Many green roof components, such as root barriers and geotextiles, are loose-laid in place over the roof surface and weighted down by the growth media. These materials are lightweight and therefore are prone to displacement by wind during the phasing of the project. A good rule of thumb is to only lay out quantities of materials that can be adequately ballasted by

Figure 5.18   Roof drain access chambers can be used with modular green roofs to provide an attractive alternative to leaving the roof drains exposed.

the end of that work day. The roofing industry has long struggled with keeping materials in place while the permanent ballast can be positioned. Discarded automobile tires can provide temporary ballast, but they can be cumbersome to convey to and from the rooftop. Small amounts of growth media strategically placed around the roof can provide temporary ballast during construction. These piles can be easily raked to uniform depths as the rest of the growth media is distributed across the project. Layering growth media along sheet-metal detailing accessories will help hold them in place while construction progresses. Irrigation components that are to be installed below the growth media must be anchored in place and protected from damage or displacement during growth media distribution. While these temporary ballasting methods help keep materials in place during the construction process, it is a good idea to minimize the duration of the use of temporary ballast by installing the full depth of growth media as soon as practical. Though it may be tempting to lay out large areas using temporary ballast so that growth media can be distributed over the entire project, sudden storms can displace materials, resulting in damage to green roof components, in addition to the cost of repositioning the materials.

Distribution of the growth media at the correct depth is critical to getting the proper coverage rates. Obviously, leveling stakes cannot be driven into the roof as is common practice for spreading soils at grade. However, PVC plumbing pipe can be cut to lengths matching the desired growth media depth, and stood on end to act as a gauge for distribution depth. Spacing the PVC pipe sections in a grid across the roof area provides leveling points on which a straight piece of lumber can be placed to maintain uniform growth media depths. Varying growth media depths can be seamlessly transitioned by incrementally positioning taller PVC leveling points.

Once the growth media has been distributed across the green roof area, one must determine where each plant must be positioned. This can be as simple as specifying random placement at a given spacing, or as detailed as plotting the location of each plant. Waving or sectioning areas to receive various species can be done using marking paint or string lines. Individual plants can be located using a variety of markers, such as colored PVC pipe sections, coffee cans, or sand bags. Erosion-prevention products, such as wind blankets or jute cloth, can significantly reduce the risk of growth-media scouring on green roof projects in coastal areas and on high-rise buildings.

All plants have sunlight and hydration requirements that must be considered when planning for delivery to the project. Shipping can place stress on the plants resulting in high mortality rates and slow establishment. Care must be taken to select freight companies and truck drivers that understand the nature of perishable plant cargo. Once plants are delivered to the project they should be immediately off-loaded to minimize stress due to heat, dehydration, and lack of sunlight. Large projects may require a staging area to be designated where plants can be stored in the sunlight and irrigated until they can be relocated to the green roof.

Irrigation

Much has been written claiming that green roofs never need to be watered; this is a gross misconception. There are succulents and drought-tolerant plants, but, there is no such thing as a drought-proof plant. Every living thing needs water to survive; some plants have greater hydration requirements than others. The challenge as a green roof designer is to design using a plant palette that affords the green roof the best chance for success and to convey realistic expectations of the watering needs of the green roof project to the people charged with caring for it. An irrigation plan that meets the needs of project must be developed. (See Fig. 5.19.)

The irrigation plan begins with providing access to water at the rooftop. Small projects less than 20 feet above the ground

Figure 5.19   Realistic expectations for hydration requirements of the plants are a critical component of every green roof design. If nature does not provide adequate rainfall to support the green roof plants, then the project owner must be prepared to provide water through supplemental irrigation.

can utilize exterior water spigots to feed water to the rooftop through a garden hose. As the height of the project increases, the head pressure required to push the water up the hose increases. Depending on the line pressure provided by the local water utility, the actual height at which water can be effectively pushed through a garden hose to the roof and then distributed to the green roof may vary. Placing a water spigot on the roof, using interior plumbing sized to adequately deliver the water at the desired pressure, will make it more convenient to irrigate, which is an important consideration when developing the irrigation plan. The more difficult it is for the caretaker of the green roof to irrigate, the less likely it becomes that the irrigation plan will be properly executed. (See Fig. 5.20.)

Once the water source has been identified or provided, one should plan for the distribution of water to the green roof plants. Here again, the irrigation requirement of the project must be determined and the cost of various distribution options

Figure 5.20   Simple planting strategies require less frequent supplemental irrigation and easily utilize garden hoses to deliver water when necessary.

Figure 5.21   A simple oscillating sprinkler and 3/4-inch garden hose are capable of irrigating 5000 ft².

must be weighed. Irrigation for smaller projects can be accomplished inexpensively using simple lawn sprinklers and garden hoses. The common lawn sprinkler can deliver 3/4 inch of water to 5000 ft² of roof area in about 45 minutes. (See Fig. 5.21.) Simple timers are available that automatically shut off the water supply after a desired duration. More complex timers are available to control irrigation cycles for more automated delivery. These are excellent for the plant-establishment period, when irrigation frequency is more regimented; ranging from irrigation every 2 or 3 days to establish prerooted plugs to once- or twice-per-day to establish cuttings or seeds. Ongoing irrigation needs typically would need to be assessed by the caretaker who could operate the system on an as needed basis.

Irrigating larger green roof projects manually or by using simple sprinklers can be very labor intensive, and a commercial irrigation system may be the only practical means of water distribution. Projects with more elaborate plant diversity may require commercial irrigation systems capable of delivering water to various areas of the green roof based on the irrigation requirements of individual plantings. The many commercial irrigation systems on the market deliver the water to the plants in one of two basic modes of delivery: drip irrigation or overhead distribution.

Drip irrigation is a general term used to describe irrigation systems that deliver small amounts of water directly to the soil at the base of the plants. This delivery method makes more efficient use of the water by reducing the amount of water lost to evaporation and runoff. Drip irrigation systems use a series of hoses with emitters that deliver water one drop at a time, allowing large areas to be irrigated with minimal water pressure. The hoses can be placed on or below the surface of the growth media and can be zoned to provide precise amounts of water to different types of plants on the same roof. Liquid-feed injection systems can be integrated with drip irrigation to deliver correct dosages of nutrients with less chemicals leaching from the green roof. Growth media formulations that use course gradations, where finer gradations allow water to move laterally through capillary action throughout the green roof root zone, may not be well suited for drip irrigation as water may travel quickly downward through the growth media.

Overhead distribution refers to the use of sprinklers that spray water through the air to deliver water to the plants. (See Fig. 5.22.) Although this method loses some water to the atmosphere through evaporation, it helps to cool the roof surface as that evaporation occurs. Growth media made up of course gradations may require this method of irrigation to evenly distribute water throughout the green roof. The higher head pressure required to operate the sprinklers may require the roof to be sectioned off into zones that deliver water to a limited number of sprinklers. Water can be delivered alternately, zone by zone, until the entire green roof has been irrigated.

Both of these kinds of irrigation systems can be completely automated based on frequency and duration of operation. (See Fig. 5.23.) There are more sophisticated systems that operate based on the moisture content of the growth media. However, many growth media formulations are too porous to allow these electrical sensors to function properly. Before specifying the

Figure 5.22 Larger projects may require irrigation systems capable of distributing water further from the source and over wider areas. This irrigation system uses hard PVC piping, commercial-grade sprinkler heads, and timer-control valves to provide irrigation for 96,000 ft^2 of green roof plants.

Figure 5.23 Sophisticated commercial irrigation systems can automate irrigation, simplifying the task for the green roof owner, but can also be used to tailor the irrigation to specific requirements of varying green roof plants, expanding the plant palette and providing greater diversity.

use of these sensors, send a sample of the blended growth media to the manufacturer for compatibility testing. To help take the guess work out of green roof irrigation requirements, there is a service available through Green Roof Blocks (www.greenroofblocks.com) called the Plant Health Alert System. The system monitors weather conditions at the green roof location and alerts the caretaker via email when heat and rainfall criteria fall outside of set limits for the particular project. This service is available through annual subscriptions for any green roof project in the United States.

6

Plants

S ome of the nation's leading experts on green roofs have recently written comprehensive manuals on green roof plants. As the green roof plant selection process is far too complex to adequately cover within this chapter, and given that other works provide such a wonderful and complete resource for green roof plant information, we won't delve into specific species and cultivars. Rather, we will discuss several plant-related issues that impact the success of our green roof project. (See Fig. 6.1.)

Native or Not

Somehow, along the path to bring green roofs to the United States, the green roof concept has been adopted by many ecologists in the landscaping industry and in the green

Figure 6.1   Green roof plant selection often requires some experimentation to identify conditions and components best suited to various plant species. The green roof in this photo is used to evaluate various growth media blends and plant species.

building movement. They have seized on green roofs as part of their prairie restoration and restorative landscaping efforts. While in theory it may appear to make sense to recapture the native landscape that was sacrificed by the building footprint, in reality the rooftop microclimate presents opposing concerns that can make this very difficult.

One factor is the growth media requirement for green roofs in the United States. Structural loading limitations require the construction of green roofs using shallow depths of lightweight, expanded-aggregate-based engineered growth media with high mineral concentration. Next, one needs to consider the climate on the rooftop; and the searing heat plants must endure. These two conditions limit the type of plants that can survive on a green roof project. However, more than growth media depth and heat tolerance, the greatest limiting factor in plant selection is the irrigation requirement. Sedums, other succulents, and ground covers can survive with little water and endure high heat. In contrast, native grasses, prairie plants, and wetland plants need aggressive irrigation regimes to survive in the searing heat of the rooftop environment.

This is particularly difficult to reconcile with the green building movement. The United States Green Building Council's (USGBC) LEED rating system awards credits toward certification for restoration of previously developed land using native vegetation to registered projects. However, the USGBC also awards credits for not installing permanent irrigation systems. The use of native vegetation is meant to reduce the use of resources consumed to irrigate and maintain typical landscaping schemes. Typically natives are only irrigated during the establishment period and the vegetation is periodically burned off to remove excess biomass, a process meant to mimic wildfires in nature. Since conducting controlled burns on rooftops is not a feasible means of reducing biomass, native plants will require routine trimming and removal of the clippings. Furthermore, several studies have indicated that permanent irrigation systems are required to provide regimeed irrigation to keep native plants alive. The continuing maintenance required by native plants in the rooftop environment puts them at odds with the intent of the native restoration credits and the principles of conservation of resources guiding the green building movement.

Does all this mean that natives cannot be used nor restorative wetlands created on green roof projects? Absolutely not; one simply must identify and understand the requirements of the various plants survival. LEED-registered projects may loose the ability to capture water resource credits due to the irrigation requirement. Growth media depths may need to be increased to accommodate some native species. More robust root barriers may be required to keep aggressive roots of some native species from damaging the roofing membrane. Dr. David Beatty, a pioneer in the U.S. green roofing concept, stated in an address to the Green Roof Congress in

Stuttgart, Germany, that "the problem with using natives in green roofs is that there is nothing native about the rooftop environment."*

Planting Strategy

In much of the United States, winter steadily has been shortened by longer autumns and earlier springs. Hardiness zones have been gradually pushing northward as the average winter temperatures have crept upward. One advantage is that early springs and late first frosts expand the planting window for green roof projects. However, perhaps the only predictable characteristic of weather is that weather is unpredictable. A late cold snap or an early frost could spell disaster for a newly planted green roof. When pushing the window with an early spring planting, it may be a good idea to specify dormant or semidormant plants. Planting after October 15 presents a risk that increases with the size and scope of the project.

Selecting plants based on winter hardiness zones can be tricky because winter hardiness is only half the story. Heat tolerance, or the lack thereof, is more likely to affect plant survivability. Plants with questionable heat tolerance should be used sparingly or in areas of the project that provide some afternoon shade. As the number of green roof projects continues to increase, experimentation will increase the green roof plant palette for various regions. Planting from a core group of groundcovers with good heat and winter hardiness will provide both visually pleasing and culturally sound base growth. Then one can spice things up with colorful annuals and natives.

Diversity

Planting a green roof with a single species may create a lush carpet of vegetation with uniform height, color, and texture. This "lawn" appearance, although visually pleasing, can place

---

*Dr. David Beatty addressing the International Green Roof Congress in Stuttgart, Germany, 2004

the green roof project at risk. Climate conditions and pests that are fatal to specific species can target a monoculture green roof and wipe out every plant on the project. Monocrop planting schemes can thrive for many years and then fall victim to catastrophic elements without warning. To limit a project's exposure to catastrophic plant failure, one should instead propagate using a diverse group of plants consisting of five or more different species. (See Fig. 6.2.)

Blending some evergreen plants into the planting scheme will give green roof projects in northern climates winter beauty. Choosing plants with varying leaf structure will produce a variety of textures while achieving greater sustainability by utilizing plants that metabolize water differently under varying conditions. Leafy plants like *Sedum kamtschaticum* store large amounts of water for superior drought and heat

Figure 6.2   Blending plant species in random planting patterns establishes a diverse ecosystem capable of sustaining weather and pest anomalies that target specific species. Blending evergreen and flowering species ensures year-round beauty of the green roof.

tolerance, but go into dormancy for the colder months and grow new leaves each spring. Plants with needle-like leaves like *Sedum sexsangulare, Sedum reflexum,* and *Sedum album* either remain bright green or turn reddish and burnt orange throughout the cold season. *Sedum spurium* has leaves that form small rosettes and *Sedum weihenstephaner gold* has small bushy leaves; both plants' leaves remain intact throughout the cold weather and turn reddish in color until spring. Including diverse blends of these species will give the green roof project the best opportunity for success and provide year-round beauty.

Seeds, Cuttings, or Plugs

There are three basic propagation strategies: seeds, cuttings, or plugs. Propagating with seed may at some point allow large numbers of plants to be inexpensively germinated over large areas. However, to date there are no green roofs in the United States that have been propagated from just seed. Seed germinating has a very high irrigation requirement; some species require constant wetting. This intense establishment irrigation requirement can be difficult to meet for green roof projects larger than 20,000 ft^2, as keeping large areas moist will likely require the installation of a commercial-grade irrigation system. Wind blankets, jute cloth, or liquid tackifier (products that bind the growth media together and biodegrade over time) may help to keep the wind and birds from displacing seed and also help retain moisture for germination. Sedum seeds are quite small and make regulating the plant spacing, planting schemes, and even distribution very difficult, if not impossible. Seeds are often sold by species in batches that can contain multiple cultivars. For example, *sedum album* seeds may produce *Sedum album murale, Sedum album coral carpet*, and other cultivars from the *sedum album* species in the same batch of seed.

Cuttings are small sections of plant material that are taken from a mature plant; this process is also referred to as cloning. The "cut end" of the cutting is placed in potting soil to grow roots and to form a new mature plant. Many sedum species readily root by simply coming into contact with the soil surface. Fresh cuttings have a finite shelf life, requiring that they are transported quickly to the project and planted. Cuttings are typically broadcast over saturated growth media to establish full coverage within the first or second growing seasons. Cuttings require irrigation during the establishment period in most regions across the United States, although some northern climates may not require supplemental irrigation for spring plantings as rainfall amounts may be adequate for the cuttings to root into the growth media. Planting with cuttings is less labor-intensive and therefore less expensive than planting with plugs. Cuttings are typically sold by the pound and in multi-species blends. This typically affords less control of plant spacing and positioning, and plant selection is usually limited to sedums.

Plugs are young plants that have developed root structures. Because prerooted plugs provide more dependable establishment of the vegetation, most green roofs are propagated using plugs. Some green roof designers use a combination of cuttings and plugs; planting the plugs initially and then augmenting in areas where vegetation is slow to establish using cuttings. Plugs typically require 6 to 8 weeks production time in greenhouses or nurseries. This production expense significantly increases the cost of planting using plugs over that of using cuttings. In addition to the added expense of the plug itself, the labor cost to physically plant plugs one at a time into the growth media is significantly more that that of planting with cuttings that are scattered across the surface of the green roof and allowed to root into the growth media. However, planting with plugs produces a viable plant over 90% of the time, whereas about half of sown seeds produce a viable plant and less than 70% of cuttings produce a viable plant. Plugs can also be precisely positioned to achieve well-defined lines between species and produce green roofs with exact quantities of varying species.

Plugs are propagated in trays that contain various numbers and sizes of individual cells. The cell size, and thus the finished plug size, varies from grower to grower. Many green roof designers mistakenly specify oversized plant stock for green roof applications. Often we will see plants specified in quart- or gallon-sized containers. This practice serves the landscape industry well in producing landscapes that rapidly achieve the design vision. Green roofs, however, have other characteristics that need consideration; primarily, the growth media. From a practical standpoint, planting 7-inch-deep potted plants in 4-inch-deep growth media would require the lower half of the potted-plant roots to be removed from each plant prior to planting in the green roof growth media. Next, consider the exacting ratios of organic to mineral blending of our formulated growth media. Introducing large amounts of organic potting soil significantly alters the water retention, weight, and draining characteristics of our growth media and can have negative consequences for our green roof. In cases where the potted material is transplanted intact into the growth media, the two materials do not blend together well. When the organic rich potting soil shrinks, voids between the potting soil and the growth media can form, exposing the plant roots within the potting soil to weather elements. Finally, the plant roots are bound by the plug cell until transplanting, when the roots rapidly spread to nourish the plant. The objective is to quickly establish the plants in the growth media to stabilize the media. Therefore, it is most advantageous to use the smallest possible plug size in order to get the plant roots established beyond the potting soil and into the growth media as rapidly as possible.

The Establishment Period

Whether one uses cuttings or plugs, there is an establishment period during which the plants root into the growth media and acclimate to the rooftop environment. This period is typically 6 to 10 weeks for plugs and about twice as long for cuttings. During this fragile stage of green roof development, the plants have high hydration and nutrition requirements. An application of a slow-release (12- to 14-month formula) granular fertilizer to the surface of the growth media at each plant will provide the nutrition the plants need during their first year. Depending on the region and the time of year, necessary hydration may be provided by natural rainfall. In cooler temperatures, rainfall or irrigation once per week is sufficient. As the weather warms, it may be necessary to irrigate to the plants with 3/4 inch of water two or three times per week. A typical lawn sprinkler can distribute 3/4 inch of water in about 45 minutes. It is best to irrigate new green roof plants during early morning hours to avoid wet foliage conditions overnight that could promote mildew and disease.

The success of a green roof project hangs in the balance of meeting the needs of the plants during this critical phase. Properly established plants are better prepared to withstand drought and winter freezing for years to come. The better care the plants receive during the establishment period, the more rapid the roof can achieve plant coverage. Aside from allowing people to more quickly enjoy the aesthetic benefits of the green roof, rapid plant coverage is important for several other reasons. The mature plant canopy will shade the growth media and help retain moisture to hydrate the plants. A dense plant canopy does not leave a lot of room for invasive weeds to find their way into the growth media. A mature plant community will more effectively transpire storm water and cool the rooftop.

Drought Tolerance

Drought tolerance is a broad term used to signify a plant's ability to withstand a period of time without hydration. This is an oversimplification of a complex characteristic of the plants selected for a green roof project. Drought tolerance and heat are very closely related; as heat rises, the length of time a plant can survive without hydration shortens. This makes drought tolerance vary for plant species from region to region. In some cases, the survivability of a plant on a green roof can vary dramatically across regions; even among those in relative close proximity. *Sedum acre*, for instance, is a drought-tolerant plant that is widely used in the Chicago area, but performs poorly in green roofs in the St. Louis area. To further complicate the plant-selection process, the climate is always changing. For example, 2006 brought cooler and wetter weather to the Chicago area; this was a welcome change for the green roof industry because the hot, dry weather of 2005 claimed many green roof plants. To ensure the success of a green roof project, one must identify the plants that have higher hydration requirements and the areas of the rooftop that may amplify these requirements. We then need to develop an irrigation plan to provide water based on the specific requirements of the various plants in the various areas of our rooftop. It is important not to confuse the term "drought tolerance" with "drought proof." Every plant needs water to survive, so there is no such thing as a "drought-proof" plant. The best preparation for drought is to select hardy plants and to put into effect a contingency irrigation plan.

Fertilizer

Green roofs generally require fertilizer over the entire planted area through the first 3 to 5 years. Beyond 5 years fertilizer use may be discontinued or localized on an as needed basis.* Several commercial fertilizers are available from horticultural suppliers. Slow-release granular formulations keep the amount of nutrients leaching from the green roof to a minimum. Popular brand names like Osmocote and Nutricote are coated pellets that have heat-activated, slow-release patterns that feed the plants for an entire year with a single application. When temperatures rise above certain levels, these fertilizers release nutrients that are carried into the root zone by rainfall or irrigation water. It is possible, under certain conditions, for these fertilizers to damage the plants. For example, nutrients can be released and allowed to accumulate on the surface of the growth media during hot and dry conditions. Rainfall typically flushes these nutrients through the green roof system to feed the plants. However, trace amounts of rainfall following several weeks of nutrient loading can carry high concentrations of fertilizer into the root zone, burning the roots and harming or killing the plants. Care must be taken when using these products to thoroughly irrigate after these extended dry spells. (See Fig. 6.3.)

There is a fertilizer called IBDU that is water soluble. IBDU granules slowly dissolve to deliver nitrogen to the plant roots.

---

*FAQ-www.greenroofs.com, 2007

Figure 6.3 Heat released by fertilizer damaged these green roof plants when a 1/8-inch rainfall in July followed a 6-week drought. Nutrients released at the surface of the growth media during the warm weather for 6 weeks were delivered in high concentration to the plant roots by the light rainfall. Providing supplemental irrigation equal to 3/4 inch or rainfall after the third or fourth week could have prevented this damage.

Large granules are available that have the ability to feed the plants for 12 to 14 months. Although IBDU dissolves very slowly, there is the potential to leach nitrates from the green roof during extended rainy periods. Blending IBDU with Osmocote or Nutricote can provide balanced nutrition and a more reliable delivery regime, an especially effective strategy during the establishment of the green roof plants. Green roof projects with irrigation systems can use liquid fertilizer delivered through the irrigation system. Although injecting liquid fertilizer is more complex than simply broadcasting granules over the roof, liquid injection systems allow exacting fertilizer regimes to be delivered to different sections of the roof to meet the nutrition requirements of specific plants.

Carnegie Mellon University, Pittsburgh, Pennsylvania.

City Hall, Atlanta, Georgia.

Children's Hospital (summer), St. Louis, Missouri.

Children's Hospital (winter), St. Louis, Missouri.

City Hall, Seattle, Washington (photo courtesy of Linda Velazquez).

Eastern Village Co-housing, Silver Spring, Maryland.

Cook + Fox offices, New York, New York.

Elementary school, Unterensingen, Germany.

Garage roof, Kansas City, Kansas.

Kerr Foundation, St. Louis, Missouri.

Mong Ha Fort, Macau, China.

Museum, Köngen, Germany.

Renaissance Hotel, Karlshrue, Germany.

World Trade Center, Boston, Massachusetts.

Zinco Headquarters, Unterensingen, Germany.

7

Maintenance

As their popularity has drawn them into the mainstream media, much has been written about green roofs during the last few years. Professionals in the green roof industry enthusiastically grant interviews, hoping that the exposure will advance the green roof concept and at the same time promote individual products. Unfortunately, the final copy is often a less than accurate account of the interview, including statements like "never needs watering" and "zero maintenance required." While the writer's challenges are to produce attention-grabbing headlines and to write articles that keep the reader interested, it has placed a huge burden on green roof professionals to debunk these misstatements and to provide architects, building owners, and green roof caretakers with the realistic maintenance requirements of green roofs.

Hydration

As stated previously, there is no such thing as a plant that never needs water because every living thing needs water for survival. An important aspect of good green roof design is convenient access to water on the rooftop. Getting the water to the rooftop is only half of the task; the other half is to set good guidelines on when to water the plants. Since most green roof projects will utilize sedums as the core group of plants, the discussion here will concentrate on the hydration requirements of this plant family. However, the hydration requirements of any herbaceous plants included in the planting scheme will need to be identified as well. Most sedums thrive in well-drained soils that keep the roots from sitting in water, which promotes root disease fatal to the plant. Sedums are easier to kill by overwatering than by underwatering.

The watering regime begins with the propagation of the green roof. Many green roof professionals prefer to saturate the soil prior to planting cuttings and plugs. This practice helps keep fresh cuttings hydrated during application, when they are susceptible to drying out under direct sunlight. When planting plugs, presaturating the green roof helps to bind the growth media for dibbling (using a tool to make impressions in the surface of the growth media) making insertion of the plugs easier, increasing the speed at which the roof can be planted. (See Fig. 7.1) Once cuttings are distributed across the green roof, wetting the tackifier or wind blanket to keep the fresh cuttings moist will likely require visual inspection because weather conditions will vary irrigation intervals. To minimize stress to the plants from transporting and handling and to promote rapid growth, freshly planted plugs need to be watered thoroughly by saturating the growth media the day the plugs are planted.

Keeping the sedums hydrated during the period following planting will help encourage rapid plant establishment and enhance the long-term success of the green roof project. Again, the interval between irrigations for cuttings is somewhat shorter than the interval for plugs. Cuttings need to be kept moist while they root into the growth media. Once rooted,

Figure 7.1   Prewetting the growth media helps bind the material so that inden-
tations can be made in the surface of the growth media, making the insertion of
plugs easier and faster.

however, they essentially become small plugs and their hydration requirement lessens. Newly planted plugs need to be
watered every 3 or 4 days during the 6- to 8-week establishment period. Once established, the watering frequency should
be gradually reduced, weaning the plants off of artificial irrigation entirely.

Established green roofs thrive most of the year on rainfall
alone. However, the green roof will require supplemental
irrigation for occasional periods without rainfall that exceed
4 to 6 weeks. When high heat accompanies these dry spells,
irrigation may be required more frequently. For most regions
of the country, supplemental irrigation may only be required
on a couple of occasions per year. In arid climates like Las
Vegas and Phoenix, a regular irrigation regime may be
necessary.

Fertilizers

A slow-release granular fertilizer with a 12- to 14-month release formulation will feed green roof plants for 9 to 10 months. (See Fig. 7.2.) The heat-triggered release rate is accelerated when these fertilizers are used in the rooftop environment. An early-spring application will provide continuous feeding throughout the growing months. Annual fertilization of the green roof is necessary for the first 3 to 5 years. Once the green roof ecosystem is established, fertilizing can be conducted locally to meet the individual needs of certain plants or over the entire green roof every few years to maintain the desired color and level of robust growth.

Rapid-release and natural fertilizers like manure can leach harmful contaminants into the drainage system and storm-water

Figure 7.2   Slow-release fertilizers are formulated to have specific release durations. As noted on the bag, the manufacturer has designed this fertilizer to provide nutrients for 12 to 14 months.

sewage system.* Unlike working with plants at grade, where the earth acts as a grand filtration system, the green roof is a shallow artificial planting surface. Please remember that elements introduced into the green roof have the potential to enter the roof drain or downspouts in high concentrations. Most people working with green roofs are motivated by environmental consciousness. The very last thing they want is to allow a green roof project to contaminate the very environment they are attempting to restore or benefit.

Weed or Not?

Some of green roof caretakers have genuine expertise in plant gener while others have difficulty differentiating a weed from a desired green roof plant. Some weeds look like weeds; long, straggly, and out of place. Other weeds, like *spurge* (*Euphorbia*), actually have pleasant blooms and a low-lying profile that blends into the green roof plant scheme. (See Fig. 7.3.) Pleasant-looking or not, weeds can compete for water and food and choke out the intended green roof plants. Weeds can also have much more aggressive roots that, when left unchecked, are capable of exploiting weaknesses in root barriers and damaging roofing membranes. Weeds can be introduced into the green roof after construction or can be brought to the project within the growth media. Proper handling of the growth media can eliminate many weed headaches early on, but some weeds are bound to find their way into the green roof sooner or later. Many of the weeds that sprout during the wet spring season will die out once the summer weather turns hot and dry. Weeds that are still alive in the fall pose a threat to the green roof system and should be removed.

While many people love the appearance of lush green roofs, few enjoy pulling weeds to keep a green roof looking desirable. Unfortunately, there is no magic bullet that can eliminate weeds from a project. Weed eradication usually involves garden gloves and energetic weed pullers. The use of

---

*__John Howell,__ *"Topdressing and Sidedressing Nitrogen,"*, Soils Basics-Part V, **Department of Plant and Soil Sciences,** *University of Massachusetts, Amherst Extension, 2007*

Figure 7.3   Some weeds are more obvious than others. While this prostrate spurge weed may be more attractive than the sapling that is rooted next to it, both will compete with the sedums for nutrients and sunlight. Therefore it is important to control weeds early in order to give the desired green roof plants every opportunity to thrive and the green roof project every opportunity for success.

herbicides is strongly discouraged due to the same environmental concerns mentioned in the fertilizer section, and so it bears repeating: unlike working with plants at grade where the earth acts as a grand filtration system, the green roof is a shallow artificial planting surface. Please remember that elements introduced into a green roof have the potential to enter the roof drain or downspouts in high concentrations.

Pests

Green roof plants can be susceptible to unwanted guests from time to time. Curious crows like to pull out freshly planted plugs to see what has been hidden in the growth media. Unfortunately, they do not put the plugs back once they have satisfied their curiosity. Plugs left out of the growth media for too long will dry out and die. There are a variety of methods used to ward off unwanted birds, such as mounting plastic owls or rubber snakes. This tends to be less of a problem as the green roof matures and the plants are well rooted. Occasionally, some larger birds on the protected species list may nest in the green roof. Persuading these birds to seek accommodation elsewhere can be a delicate issue because gaming laws and wildlife protection ordinances may limit the disruption of the natural migration and mating of certain species. Often human activity on and around the green roof is enough to convince these visitors that there may be quieter settings for them to explore.

Other pests can come in much smaller forms: aphids and other plant-eating insects. Routine inspection of the plant foliage for these pests will help identify the problem before it gets out of hand. Natural and environmental friendly means of control should be the primary course of action. The use of pesticides should be reserved as a last resort and then used with the utmost caution. It bears repeating once more: please remember that elements introduced into a green roof environment have the potential to enter the roof drain or downspouts in high concentrations.

Survival of the Fittest

Often one has a distinct vision of how he or she wants the green roof to look, what plant species will be included, and where they will be located. Nature, however, has its own vision and usually imposes its will despite humankind's best efforts. Dominant species will edge out weaker or slower growing species. As a result, green roofs that are propagated with six or more species may only contain a few of these species after 5 years. Furthermore, continually changing climate conditions may act to further pare down species diversity over subsequent years, allowing the reemergence of earlier species, or even permitting an entirely new plant scheme to develop. Maintenance plans should include an inspection of the green roof project in early spring following the first winter to identify and replace any plugs that did not survive the winter. Although winter mortality rates should decrease as the green roof matures, replenishing missing species annually will help maintain plant diversity.

After several years, it may become apparent that conditions on the green roof are not conducive for the survival of a certain species. One may also find that some species do better in some areas of the project than in other areas. It may be necessary to experiment to find out which species thrive as well as which areas are best suited to various species. As the conditions on the green roof change from time to time, the maintenance regime must change to keep reflecting these changing conditions. Observing and participating in the evolution of the green roof ecosystem can become a rewarding experience. While some green roof projects may require more attention than others, no green roof project will be completely maintenance-free.

8

Realistic

Expectations

The goal for green roof designers is for those enjoying, caring for, and paying for the green roof to be completely satisfied with the time, effort, and money they have invested. The first and most critical step toward success in this endeavor is to convey realistic performance expectations from the outset. People tend to be more understanding about the less-than-attractive aspects of the green roof development process when they have been properly informed beforehand. When armed with the necessary information to answer questions posed by other stakeholders, green roof design team will not find themselves in the hot seat when the green roof plants do not mature overnight, or when other issues or complaints arise. Rather, they will shine as knowledgeable sources of green roof information while they educate these inquisitive individuals on the particulars of their green roof. It is through the sharing of such information that even the most skeptical maintenance technician is transformed into a green roof enthusiast, and more often, a big fan of the green roof designer.

Coverage

One of the most common conflicts between green roof designers and those parting with precious dollars to pay for the green roof is the amount of time it takes to achieve complete plant coverage of the growth media. Coverage time can vary based on the species and the spacing of the green roof plants. Typical spacing of one plant per square foot generally achieves full coverage within the first 2 or 3 years from the planting date. Increasing the planting density to one plant every 8 inches on center will decrease the amount of time it takes for all of the bare spaces to fill in with plants, but will add significantly to the planting cost of the green roof project. It may be necessary to plant slower growing species like *Sedum sexsangulare* at increased density, while quicker growing species like *Sedum spurium* may be planted at the one plant per square foot density. Varying the planting density for different species used on the same roof will allow the plants to mature and achieve coverage more uniformly.

Seasonal Appearance

As green roofs are ever-changing ecosystems, their appearances can change from season to season and from year to year. Some plants thrive in hot, dry summer conditions and go dormant for the cold winter; this characteristic helps these plants survive. However, dormant plants are often not very attractive during the winter season. A blend of species to include some evergreen plants will help give the green roof some winter beauty. Some species like *Sedum acre* look great in the spring, but show signs of heat stress during the hot summer months. Once cooler fall temperatures replace the summer heat, the *Sedum acre* recovers and lush green growth returns. These seasonal appearance variations are a part of the natural processes taking place on the green roof. Conversations about these appearance variations early in the design phase will go a long way to convey realistic expectations of how the green roof project may look during different times of the year.

Irrigation

Some newspaper and magazine articles make erroneous claims about never having to water green roofs. Therefore, it is important to clearly communicate irrigation needs with those involved in a green roof project early. The time to discuss the irrigation requirements of green roof plants is long before cutting the ribbon at the dedication ceremony. Green roofs in southern climates may not survive without some supplemental irrigation. These are discussions that must take place early in the design phase of any green roof project. Rainwater-harvesting strategies can help ease the irrigation demand on potable water sources. Once again, a written maintenance plan with clear irrigation instructions will eliminate any misunderstandings regarding the irrigation requirements of the green roof project.

Manicuring

Some green roof projects will leave decisions about how the plant species propagate the rooftop up to nature. The natural ebb and flow of the ecosystem in those projects will determine if some species dominate others. However, other projects will be designed with well-defined transitions between areas planted with different species. As nature's forces begin to blur these lines of separation, a decision will need to be made as to how forcefully one is going to maintain these lines. This can include pulling out dominant plants and repopulating weaker species with new plugs or cuttings. Additionally, some species may spill out onto walkways and no-vegetation zones. Care must be taken when trimming these plants not to allow the cuttings to propagate in unwanted areas of the roof. More complex green roof projects that include water features, sculptures, and sitting areas will likely require a more aggressive manicuring

Figure 8.1   Notice how one of the green roof plant species has jumped over the boarders into the walkway and into the sectioned-off serenity area. Many green roof projects begin with well-defined boundaries and lines of separation between different plant species only to have nature blur or eliminate these lines throughout the evolution of the maturing green roof.

regime that adds to the maintenance cost. These costs need to be identified early so a long-term commitment can be made to perform the grooming and manicuring required to preserve the original design vision. (See Fig. 8.1.)

Annuals

Annuals can offer a splash of color and diversity to a green roof project. Often annuals may be incorporated into the first growing season to detract attention from bare spots during the establishment period. However, when they fail to show up the following year, they can be missed. If one uses annuals temporarily, he or she should discuss this strategy so that everyone

Figure 8.2 The use of annuals adds seasonal color to the green roof and enhances the enjoyment for the visitors.

involved realizes the annuals will not be back the next year. If annuals are to be part of the permanent planting scheme, their use must be clearly defined in the written maintenance plan. Designating areas for the use of annuals is a good way to get visitors to interact with the green roof. (See Fig. 8.2.)

Lifespan

Some plant materials and all roofing materials have a finite lifespan. Potted plants will need to be replaced from time to time. Shrubs and trees may outgrow the rooftop setting and need to be replaced with smaller plant stock. Roof flashings are typically the first part of the roofing system to deteriorate and often can be replaced to extend the life of the roofing system. At some point, the entire roofing system will need to be replaced. If a modular green roof system has been employed to host the plants, they can be easily moved to replace the roofing system. If a built-in-place green roof system has been constructed, moving and reusing plant stock can be significantly more difficult. While no one has a crystal ball to forecast when each of the green roof components will reach the end of its service life, the maintenance plan should include routine inspections to identify problem areas and early signs that replacement of the green roof component may be eminent.

9

The
Environment

The green roof concept holds great promise of helping to remediate a host of environmental issues. However, in order to make a meaningful impact on the following environmental issues, green roofs will have to become the norm rather than the oddity. At present green roof projects are sporadic; they mainly adorn a few highly visible university rooftops and some LEED-certified buildings, and are found in the minimum-square-footage amounts necessary to meet governmental mandates in Chicago, Portland, and Washington, D.C. The motivation for investing in a green roof can be broken into three categories: compliance, amenity, and research. While green roofs continue to become more commonplace, they are still somewhat of an anomaly within the construction industry. It may require some heavy-handed governmental regulations to compel widespread green roof usage in numbers necessary to substantially lower the cost. Lower cost will carry demand beyond that of mandate compliance to that of a desired amenity and of a sound investment with a more rapid rate of return.

Heat Island

The plant life that once helped cool cities has been destroyed and replaced with buildings and pavement that absorb the sun's energy and hold the radiant heat, creating concentrations of heat that now surround these densely populated areas to form islands of radiant heat. While little can be done to restore the green space that once grew where pavement now covers the ground, several technological advances are being employed to reduce the amount of solar energy absorbed by rooftops. Highly reflective materials such as Cool Color metal roofing, white membrane roofing, and roofing tiles and shingles covered with ceramic chips reflect the solar radiation instead of allowing that energy to heat the roof surface. These reflective materials result in rooftops that are 15 to 40% cooler than rooftops covered with traditional roofing materials.[*] Green roofs go beyond reflectivity that keeps the roof surface cool. The green roof plants cool the air around the rooftop through the phenomenon of evapotranspiration, the process that cools the surface of an area as water evaporates from it; similar to the way the human body cools through perspiration. Researchers estimate that constructing green roofs on 50 to 60% of the rooftops in densely populated cities could result in lower summertime temperatures by as much as 10°F.[†] Cooler cities during periods of peak demand for electricity would ease demand on electrical grid systems and help reduce carbon emissions from coal-fired power plants.

---

[*]Green Roof Environmental Evaluation Network, Southern Illinois University Edwardsville, Dr. William Retzlaff et al., 2007.
[†]*A Guide to Rooftop Gardening*, Chicago Department of Environment.

Storm Water Management

Cities like Washington, D.C., have realized the value of green roofs over traditional storm water management strategies for new developments and construction projects. Rather than routing storm water from rooftops into underground cisterns to buffer the first flush of storm water to regulate the flow of water into sewage treatment facilities, green roofs are being approved by the public works departments to instead retain water on the rooftop where it can be transpired back into the atmosphere. Research conducted at Southern Illinois University, Edwardsville, has shown that a green roof with 4 inches of growth media is capable of retaining over 50% of annual rainfall.[*] A green roof absorbs storm water until the growth media becomes saturated, at which point water begins to flow from the rooftop. This delay of storm water entering the sewage treatment system helps to ease the burden on the combined sewage systems operating in many American cities. Many in the green roof industry are focusing attention to the quality of storm water leaving the green roof. Growth media formulations, plant species, and fertilizers are being evaluated for stability and the ability to minimize the leaching of contaminants.[†] Researchers are also exploring the possibility of future green roofs that will be capable of sequestering pollutants from the air and rainfall.

Green Space and Habitat

Just as it is a challenge to populate the green roof with vegetation, it is equally challenging to populate the green roof with insects and animals. It can be unclear which insects and animals will be able to adapt to new rooftop ecosystems. While some green roofs are inhabited by insects, bees, and birds, it is not yet clear which particular species are likely to thrive in the extreme environment and which will be beneficial to the new

[*]Forrester et al. 2007.
[†]Berghage 2008; Simmons 2008; Retzlaff et al. 2008.

ecosystem. As people gain a better understanding of how flora and fauna interact in shallow green roof habitats, they can begin to design green roofs to enhance this interaction. Perhaps by varying growth media depths and including water features, green roofs can move beyond simply replacing lost green space to take on broader roles of habitat restoration and biodiversity enrichment. Despite people's best efforts to influence rooftop ecosystems, nature often has a greater influence through animal and bird feeding habits, migratory patterns, and the weather. One thing is certain, however—urban sprawl is gobbling up natural habitat at an alarming rate. Over the last two decades, over 400,000 acres of farmland have been lost to urban sprawl each year.[*] Whether one designs for habitat creation or simply allows nature to run its course, green roofs offer a feasible restoration strategy.

Air Quality

While much has been written regarding green roofs and air quality, little is actually known about the impact widespread green roof usage would have on air quality. It is known that green roof plants take in carbon dioxide and convert it to oxygen through the process of photosynthesis. However, those marketing green roofs as vast oxygen-generation engines may only be telling half the story. Oxygen is produced by sunlight-driven photosystem; it is not generated by the consumption of carbon dioxide. Therefore, the source of oxygen during photosynthesis is water, not carbon dioxide. Water may be in short supply for shallow rooted green roof succulents called *CAM* plants that tend to conserve water during daylight hours. It is likely that air quality may be more impacted by utilizing green roofs to lower the temperatures around densely populated areas typically prone to smog and poor air quality. Warm upper air over large cities inhibits vertical circulation and traps nitrogen oxides (NO_x) and volatile organic compounds (VOCs), where

[*]A. Ann Sorenson, Richard P. Greene, and Karen Russ, American Farmland Trust, *Farming on the Edge* (Center for Agriculture in the Environment, Northern Illinois University), Washington, D.C., 1997, Table 7, via http://farm.fic.niu.edu/foe2.

they react with sunlight to produce smog. High concentrations of green roofs could have a combined effect of reducing carbon dioxide and lowering temperatures to improve air quality.*

The Solution

While any green roof space is a step in the right direction, one or two green roofs in a city of 2 million people or green roofs that only cover a small percentage of the rooftop in order to meet the requirements of a governmental mandate will have little effect on any of these environmental issues. With systematic greening of the roofs on all newly constructed buildings, the development of lightweight growth media that allows green roofs to be constructed on existing rooftops with limited structural capacity, and innovated irrigation strategies that afford greater plant diversity and shallower green roof depths, the green roof numbers necessary to have meaningful impact on these environmental issues can be achieved.

LEED

The United States Green Building Council (USGBC) has developed a rating system that awards designers and property owners for various levels of environmentally responsible building strategies. The 69-point rating system is broken into five categories: Sustainable Sites (SS), Water Efficiency (WE),

*Blankenship, R. E., 2002, *Molecular Mechanisms of Photosynthesis,* Blackwell Science; Campbell, N., and Reece, J., 2005, *Biology,* 7th ed., San Francisco: Benjamin Cummings; Gregory, R. P. F., 1971, *Biochemistry of Photosynthesis,* Belfast: Universities Press; Govindjee, 1975, *Bioenergetics of Photosynthesis,* New York: Academic Press; Govindjee, Beatty, J. T., Gest, H., and Allen, J. F. (eds.), 2005, "Discoveries in Photosynthesis," *Advances in Photosynthesis and Respiration,* Vol. 20, Springer; Rabinowitch, E., and Govindjee, 1969, *Photosynthesis,* New York: John Wiley & Sons, Inc.; Stern, Kingsley R., Jansky, Shelley, and Bidlack, James E., 2003, *Introductory Plant Biology,* McGraw-Hill.

Energy and Atmosphere (EA), Materials and Resources (MR), and Indoor Environmental Quality (IE). There are five additional credits available for Innovative Design (ID). By taking prescriptive steps, the project team captures credits in the various categories to reach one of four levels of achievement: Certified, Silver, Gold, and Platinum. The following is a list of credits that can apply to projects constructing green roofs. For information on possible LEED credits associated with the use of specific green roof systems or components, contact the manufacturer. For more information on the USGBC or LEED, visit the USGBC Web site, www.USGBC.org.

SS 5.1: On previously developed sites, restore a minimum of 50% of the site area by replacing impervious surfaces with native or adaptive vegetation.

SS 6.1: If existing imperviousness is greater than 50%, implement a storm water management plan that results in a 25% decrease in the rate and quantity of storm water runoff.

SS 7.2: Install a green roof for at least 50% of the roof area.

WE 1.1 and 1.2: Reduce by 50% or use no potable water for landscape irrigation.

EA 1.1 to 1.10: Reduce design energy cost compared energy cost budget for energy systems regulated by ASHRE 90.1-1999. Green roofs have been shown to reduce energy consumption up to 75%.[*]

MR 4.1 and 4.2: Use of materials with recycled content constitutes at least 5% (for credit 4.1) and 10% (for credit 4.2) of the total value of materials in the project. Many green roof materials are made of postconsumer and preconsumer recycled materials.

MR 5.1 and 5.2: Use a minimum of 10% (for credit 5.1) and 20% (for credit 5.2) of building materials and products that are manufactured regionally within a radius of 500 miles. Many green roof providers that use local greenhouses and

---

[*]A National Research Council Canada study by Karen Liu, Ph.D., evaluates green roof systems' thermal performances. *Professional Roofing Magazine*, September 2002.

nurseries to grow modules will qualify for regionally manufactured materials.

ID 1.1 to 1.4: Use innovative and unique approaches to green roof installation. Innovative modular green roof installation methods can save the money. Savings from this area of construction afford spending on sustainable building methods in other areas of the project.

10

Return on

Investment

For many green building strategies, rapid return on investment (ROI) is an important consideration for the investor. Reducing the initial cost will hasten return on investment. To fully understand ROI, one must first identify the various areas of value. Some areas of green roof value are difficult to quantify. Green Roofs for Healthy Cities, the green roof industry trade organization, has formed two committees to develop a life-cycle cost-analysis tool and an energy-modeling tool. Though these committees are working simultaneously, the energy-modeling tool is a large component of the life-cycle cost-analysis tool and must be completed before the life-cycle cost analysis can be completed.

Life Cycle

Life-cycle cost analysis encompasses a broad range of tangible and intangible benefits. This section will focus on the life cycle of the roofing system, and some of the other benefits will be discussed in the following sections. Typical roofing systems have a lifespan of 15 to 20 years when left exposed to the elements. One can assume that by installing a green roof to protect a 15-year roofing system, the lifespan of that roofing system will be extended to 60 years. The cost of ownership of typical roofing systems left exposed to the elements can be compared to this. The initial installation of the roofing system would include all materials above the roof deck, including new roof insulation. After 15 years, the roofing membrane reaches the end of its service life and will need to be replaced. This time, as long as the roof did not leak and the insulation was kept dry, new membrane can be installed over the existing roof assembly without replacing the roof insulation. This process, referred to as a layover, is much less expensive than removing and replacing the entire assembly. However, after 15 more years, the entire assembly must be removed to the roof deck and the new roofing system to be installed must now include new roof insulation. After the third roofing system reaches 15 years of service life, it will receive another layover that will provide the remaining 15 years in the 60-year lifespan comparison. When calculating the cost of each of the four roofing systems, one must estimate the rate of inflation for the future roofing work. While this estimate cannot be made with certainty, one can average recent inflation rates based on the Consumer Pricing Index. Between 1985 and 2005, the rate of inflation hovered between 2 and 4%. The following cost analysis assumes an annual rate of inflation of 3% and the cost of roofing work conducted in the Midwest at prevailing wage rates.

Initial roofing system (15-year service life)	$7.50 per square foot
First layover (15-year service life)	$7.56 per square foot
Second entire replacement (15-year service life)	$17.67 per square foot

| Second layover (15-year service life) | $18.36 per square foot |
| Roofing investment over 60 years | $51.09 per square foot |

Compare this with a green roof cost of less than $25 per square foot.

Energy Savings

There are three variables affecting energy consumption that make estimating savings contributed to green roof installation very difficult. The insulating value of the building envelope is the most significant factor impacting the energy consumed by heating and cooling the structure. The efficiency of the heating and cooling equipment is another factor that determines the amount of energy consumed by heating and cooling. The most difficult factor to consider when formulating energy consumption estimates is the weather. Climate variation from region to region dramatically influences heating and cooling energy consumption. Moreover, weather variations from year to year vary the heating and cooling requirements for the building. These variables make developing the green roof energy-modeling tool a herculean task. In addition to the work Green Roofs for Healthy Cities is engaged in, proposed research at Southern Illinois University, Edwardsville, would work toward establishing a baseline for energy savings resulting from green roof installation. The project would seek to develop coefficients by which the baseline could be factored to account for these variations. Realistically, it will likely be several years before either of these projects produces functional modeling tools for energy savings derived from green roofs.

Storm Water Fees

Many communities across the United States have adopted storm water treatment fees to fund the treatment of storm water runoff. Most of these fees are assessed across the entire population based on treatment cost rather than charge based on the impervious surface area on each property within the community. While it may be easier to include this funding in sales or property taxes, it may be a less than equitable means of assessing this type of tax, as a significant portion of the storm water treatment cost is generated by the owners of property on which there are vast areas of rooftops and pavements. Communities that shift to an assessment method based on contributory impervious surfaces will likely provide credits toward the storm water fees for areas covered with green roofs, pervious pavement, and other storm water retention strategies. Currently, storm water fees are unrealistically low. As the fees are adjusted to reflect the actual cost of storm water infrastructure and treatment, offsetting credits for green roofs will help to produce more rapid ROI for green roof investments.

Intangible Benefits

Some benefits to green roof investment are difficult, if not impossible, to quantify. The Olsen Garden on the rooftop of the seventh floor at the Children's Hospital in St. Louis, Missouri, provides patients, family, and staff with a brief distraction from the rigors of medical treatment and work within the hospital. A study concluded by Louisiana State University in 2002 concluded that medical patients recover faster when exposed to views of nature. To give some perspective on the ongoing value the administrators place on the Olsen Garden, the initial cost of the rooftop garden exceeded $1,000,000 and the annual maintenance budget exceeds $35,000. (See Fig. 10.1.)

Loft apartments with green roof access command higher leasing rates than identical apartments without green roof access. These are examples of green roof projects where the property owner has attached a dollar value to the green roof. The office of Cook + Fox, an architectural firm in Manhattan, is arranged so that most of the workstations have direct line of sight to the company's green roof. (See Fig. 10.2.) The USGBC cites studies that indicated productivity increases and absence due to illness decreases for employees who enjoy views of

Figure 10.1 The Olsen Garden is situated on the rooftop of the seventh floor of the St. Louis Children's Hospital. The rooftop garden includes a soft composite rubber pathway that winds through natural grass play areas, water features, sculptures, and a diverse selection of plants.

Figure 10.2 Work stations from all areas within the Manhattan architectural firm Cook + Fox have direct sightline views of the radius window wall overlooking the green roof.

Figure 10.3   Employees of Cook + Fox volunteered their Saturday to create this 4000 ft² green roof in Manhattan, New York. In summer 2008 the entire modular green roof was temporarily relocated to an adjacent section of the building to accommodate a complete removal and replacement of the roofing system. The green roof modules were reinstalled after the roofing work was completed; mature plants all intact.

nature in their work environment. However, this benefit is extremely difficult to quantify on paper. Furthermore, the employees at Cook + Fox volunteered to construct the green roof on a Saturday. (See Fig. 10.3.) The partners of the firm, Richard Cook and Robert Fox, have described a resulting sense of ownership of the green roof shared by their employees that exceeded any of their expectations for the project. How can one attach a dollar value to a sense of pride and accomplishment? Clearly, the benefits of green roofs extend beyond monetary and environmental enhancements. These are just a few examples of the various benefits of owning, living near, working by, and interacting with green roofs. There are as many benefits as there are green roofs.

11

Quantifying

the Benefits of

Green Roofs*

*This chapter was contributed by W. Retzlaff, S. Morgan, K. Forrester, J. Gibbs-Alley, and S. Kaufman of the Green Roof Environmental Evaluation Network at Southern Illinois University Edwardsville.

As described in previous chapters in this book, a green roof typically consists of plants in a growth media with a root barrier to prevent plant roots from penetrating the roof membrane. When it rains, water flow is intercepted as it passes through the growth media, allowing some of the rain to be absorbed by the plant. The pore spaces of the growing medium also retain storm water. These same pore spaces, along with the plants and growth media, also act as a thermal barrier for the building below. Selecting design criteria to maximize the environmental benefit is a critical step in constructing and maintaining a green roof. This chapter elaborates on some of the research that documents good design practice and the environmental benefits of green roof systems.

Storm water runoff has become a serious environmental concern not only in urban areas but in expanding suburban areas as well. As natural ecosystems and agricultural areas are turned into strip malls with vast areas of impervious surfaces, the problem will only intensify. According to the U.S. Environmental Protection Agency (USEPA), a typical city block generates more than five times as much storm water runoff as a woodlot of the same size.[*] Municipalities are struggling to keep up with increased storm water runoff as their communities expand. Some communities, like St. Louis, Missouri, and Indianapolis, Indiana, have combined storm water and sewage systems. During large rainfall events, wastewater treatment facilities are inundated with storm water and untreated sewage effluent flows into the nation's water supply.

Another environmental concern can arise because of vast areas of impervious surfaces. Temperatures in urban and suburban areas average 1 to 6°C warmer compared to surrounding rural areas. This results in formation of the "heat island" effect, which can cause the temperature to reach critical health levels. The heat wave in July 1995 in Chicago was one of the worst weather-related disasters in Illinois history, with approximately 525 deaths over a 5-day period. As noted by Changnon et al.,[†] "The loss of human life in hot spells in summer exceeds that caused by all other weather events in the United States combined, including lightning, rainstorms/floods, hurricanes, and tornadoes." Further, in cities with a population over 100,000, the heat island effect results in a 3 to 8% increase in energy demand.[‡]

This book has shown that a green roof may mitigate both of these environmental concerns as well as others. It has also shown that the green roof planning process is complex, which causes many developers, builders, or owners to wonder if it is worth it. Some fear that they might commit to a green roof and then after the first year or even in the first 6 weeks it might not be "green" either in appearance or in practice. This chapter should provide the information required to put the final pieces of the green roof puzzle together and help make some critical design choices in the final stages of the process.

---

[*]USEPA, 2003
[†]Changnon, S. A., K. E. Kunkel, and B. C. Reinke, 1996. "Impacts and Responses to the 1995 Heat Wave: A Call to Action." *Bulletin of the American Meteorological Society* **77**(7):1497–1506.
[‡]USEPA, 2007

Research—Green Roof Plants

Selecting plant species for a green roof is an important decision that may determine the viability of the green roof over the long haul. This chapter digs deep—teasing out valuable insights in the search for the plants that will work on a particular green roof. Green roof plant selection can be complicated, but there are certain characteristics in a plant that can be helpful in the unique confines of the green roof. These include drought resistance, ability to withstand extremes of heat and cold, low growing, shallow roots, and long life expectancy. Plants must also be relatively low maintenance and require little to no fertilizer input or irrigation.[*] Besides the physical characteristics, a species must also be readily available and cost effective to be a successful green roof plant. For most green roof owners, the green roof plants must also look good. Two groups of plants fit these characteristics—either native species or succulent groundcovers.

The above-listed characteristics should become a shopping list when searching for plants suitable for an extensive green roof. The *crassulaceae* family is filled with plants with succulent leaves that tend to be found in dry, arid environments where water is scarce. One member of the *crassulaceae* family, the genus *Sedum*, has emerged as one of the hardy succulents that have the tools to survive the harsh green roof environment.[†] Sedums utilize many of the survival methods that most drought- and high-temperature-tolerant plants employ, including storage of water in leaves and stems, Crassulacean acid metabolism (CAM) photosynthesis (which means that these plants fix CO_2 at night and, therefore, reduce water loss), and shallow root systems. Most sedums are low-growing succulent plants that thrive in full sun and long dry periods, but can also withstand shade and temperature extremes. Sedums are also generally long-lived and are self-propagating or rerooting plants which help make

---

[*]Dunnett, N., A. Nolan, 2004. "The Effect of Substrate Depth and Supplementary Watering on the Growth of Nine Herbaceous Perennials in a Semi-Extensive Green Roof." *Acta Horticulture,* **643**:305–309.

[†]Snodgrass, E.C. and L.L. Snodgrass, 2006. *Green Roof Plants.* Timber Press, Portland, Oreg.

them a cost-effective choice for the roof. Many sedums are also known for the ability to easily propagate and to produce quick coverage over a roof area.[*]

In addition to using sedums, there has been some desire to use native plants on a green roof. In traditional ground-level gardens, native plants are associated with lower costs due to lower input. Native plants usually do not require excessive soil preparation, fertilizers, irrigation, or pruning. Native plants are also thought to help bring native fauna to the roof, thereby increasing urban biodiversity.[†] However, there are two potential problems with using some native plants in a green roof environment. First, the green roof environment is not like that of a traditional ground level garden. The green roof faces many unique environmental challenges that are not faced in a traditional garden. The second problem lies within the type of native plants available in your region. For example, in Illinois about 55% of the state once was covered by prairies that were composed of various grasses, sedges, legumes, and members of the *compositae* family.[‡] Prairie plants typically require deeper growing medium depths (often greater than 40 centimeters in depth) than can be accommodated on an extensive green roof. Grasses also tend to attain a higher biomass, which can pose problems with structural roof loading as well as create a fire hazard.[§] Additionally, many native prairies also rely on periodic fires to maintain their natural balance, which may not be replicated on a green roof. For these reasons, many native plants have not been utilized in extensive green roof systems.

With so many different plant choices, the plant selection process for the successful green roof can be difficult. Each green roof has to be able to survive in the regional climate as well as its own specific microclimate. With these limitations, research must be evaluated from many different climates and conditions to be able to suggest the best green roof plant species for the region. Each green roof designer, installer, and

---

[*]Snodgrass and Snodgrass, 2006.
[†]Lacey, L. "Why Grow Natives?." *Growing Native Newsletter*, Berkeley, Calif. Available Online: www.stanford.edu/%7Erawlings/gronat.htm, 1994.
[‡]Historic: Illinois Country: Plants. Available Online: www. museum.state.il.us/ muslink/nat_amer/post/htmls/ic_plants.htm, 2007.
[§]Snodgrass, 2006.

owner must do their homework to avoid documented problems (many of these have occurred in research projects conducted by the Southern Illinois University, Edwardsville, Green Roof Environmental Evaluation Network, SIUE G.R.E.E.N., or at other university research centers). Using research information to guide decision making in green roof projects can save hours of exasperation at a later date—a list of suggested resources is provided at the end of this chapter.

As mentioned previously, sedums have become a popular plant choice for extensive green roof installations for several reasons. Sedums withstand high temperatures, drying winds, and periods of drought because many are succulents. Sedums have high water-use efficiency, which allows them to survive drought conditions where other plants might not.[*] Sedum plants in a 6-centimeter growth media depth could support growth with 28 days between watering.[†] This is an important benefit when irrigation on a rooftop is limited or nonexistent. A drought tolerance study in Michigan compared three *Sedum* species: *S. acre L., S. reflexum L.,* and *S. kamtschaticum ellacombianum* Fisch. along with two other non-sedum native plant species (*Coreopsis lanceolata* and *Schizachyrium scoparium* Nash). All plants were grown in 7.5 centimeters of growth media depth. All three *Sedum* species maintained active photosynthesis and survived the 4-month drought study, while the native *C. lanceolata* and *S. scoparium* Nash needed water every other day to survive.[‡] Another Michigan green roof study, using 10 centimeters of growth media, examined growth, survival rates, and visual appearance of nine *Sedum* species and 18 native plant species over a 3-year period. All *Sedum* species tested in this second Michigan study were found to be acceptable for green roofs in the upper midwest, while other non-sedum species were not.[§]

In a plant species study at SIUE, one of the first significant findings was a fertilizer injury incident (See Fig. 11.1) that occurred 1 month after the first modular green roof systems

---

[*]Gravatt and Martin, 1992.
[†]van Woert et al., 2005.
[‡]Durhman et al., 2006.
[§]Monterusso et al., 2005.

Figure 11.1   Effect of fertilizer injury under low rainfall conditions on (*a*) *Sedum floriferum* "Weihenstephaner Gold" (foliage burning) and (*b*) *S. sichotense.*

were placed up on the SIUE Engineering Building research roof.[*] In this study, Osmocote slow-release fertilizer was being used and there was a series of very warm days (greater than 90°F daytime temperatures) followed by a rainfall of less than 2 millimeter. *S. floriferum* "Weihenstephaner Gold" suffered fertilizer injury from the commercial slow-release fertilizer following a minimal rainfall [Fig. 11.1(*a*)] event while other species, like *S. sichotense*, experienced no injuries [Fig. 11.1(*b*)]. A fertilizer blend of Osmocote or Nutricote and IBDU is now recommended to avoid this injury issue.

Some green roof owners wonder if they need to fertilize their green roof plants at all. In an experiment at SIUE on the Engineering Building green roof, it was demonstrated clearly that not using fertilizer at the initial plant installation is unacceptable (See Fig. 11.2).[†] Unfertilized green roof systems in

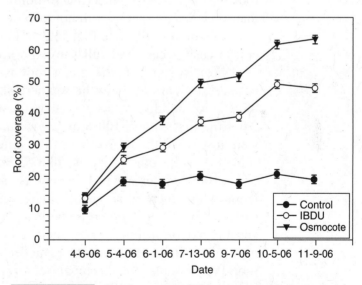

Figure 11.2   Roof coverage (%) by green roof plants during the 2006 growing season that have been fertilized by IBDU, Osmocote, or no-fertilizer (control) once in 2005 and once in 2006—3.7 grams per plant of IBDU and 5.3 grams per plant of Osmocote was applied at planting in 2005 and again once in 2006. (Error bars ±1 standard error.) (Gibbs, 2008).

---

[*]Kaufman, 2008.
[†]Gibbs, 2008.

this study had less than 20% of their surface covered by plants at the end of the second growing season. Clients would definitely not see a "green" roof if the installer did not fertilize at initial installation—an unacceptable outcome. There has been some discussion among green roof research groups that, in practice, one may not need to fertilize once the roof becomes fully established or that fertilization could occur at greater time intervals or with limited fertilizer rates. Whatever the final outcome in the fertilizer research, it is clear that sedums have some nutritional requirement and cannot be "abandoned" on the roof.

Other environmental factors in the rooftop microclimate can be critical when making plant choices and, unfortunately, these factors are often not considered. In Chap. 2, the consequences of air conditioning ducts and roof vents and their impact on a green roof were discussed. On the green roof on the SIUE Engineering Building, it was discovered that some *Sedum* species will not tolerate winter shade while others do quite well. (See Fig. 11.3.) Though that green roof enjoys easy access and full sun exposure (which makes it very hot) in summer, the green roof is shaded all winter because the sun lowers in the southern sky and the adjacent three-story building casts its shadow over the green roof. After the first winter, 100% of *S. hybridum immergrunchen* were lost when the roof was in full winter shade (no direct sunlight) while other species, like *S. sexangulare,* experienced 100% winter survival. In contrast, at a field-site installation where there is no shade, after 3 years there is 100% survival of *S. hybridum immergrunchen.* Installers are encouraged to carefully review their sun exposure and possible roof shading before making plant decisions for any green roof. For example, *S. spurium* and *S. ternatum* have survived in full shade on a roof for two growing seasons (these plants are growing on the roof in an area that does not receive direct sunlight; winter or summer).*

---

*Hise et al, 2007.

Figure 11.3   Effect of full winter shading on survival of (*a*) *Sedum hybridum immergrunchen* (no plant survival after one winter in the shade) and (*b*) *S. sexangulare* (100% plant survival) on the SIUE Engineering Building research green roof.

Research—Green Roof Growth Media

The physical and chemical properties of the growth media in a green roof may also affect plant viability, roof coverage, and plant growth. The ratio of inorganic material to organic material in the growth media can determine whether the plants grow and survive or perish and leave the owner with a nongreen roof. Imagine going through the entire planning process followed by a successful installation and then losing the green roof plants after the first 6 weeks or 6 months. On a research green roof, this is not a problem—one just asks more questions, develops a hypothesis, and establishes more projects. On a commercial installation, the first and cheapest solution may be to replant either with the same plant selections or some new plant selections. However, consider this scenario: in about 6 or more months failure occurs once again—all the plants on the green roof once again are dead. It might not be the plants. The components of the growth media provide critical nutrients and water holding capacity for the plants and also provide some of the environmental benefit of the green roof—retaining storm water, reflecting sunlight, and easing the thermal flux.

One of the first decisions about growth media (which can determine many growth media decisions) is the weight. In Chap. 2, weights and measures were explained very carefully and that information can indicate the proper weight that a roof structure will support. At this point, a growth media composition that meets the weight criteria and also provides the nutrition and water necessary for good plant health and survival is needed.

The composition and depth of the growth media can have a major impact on plant selection. Traditional soils are typically too heavy, especially when wet, for use on a rooftop so a different type of growing substrate has been developed after much trial and error. This green roof growing substrate is referred to as "engineered growth media." Sherman* suggests that at least four factors are important when determining if a growth media is

---

*Sherman, 2005.

viable: water-holding capacity, degree of drainage, fertility for vegetation, and density. These factors also determine some of the environmental benefits of the green roof. The growth media must be able to resist heat, frost, shrinkage, and sparks—all factors damaging to normal roofs. In most cases, growth media contains large amounts of inorganic (lightweight) material and small amounts of organic compost. The actual composition depends on the specifics of the green roof. The Chicago Department of the Environment recommends that growth media be lightweight and made of high-quality compost and recycled materials.[*] In Atlanta, one firm is using growth media that is primarily sand-based with expanded clay or slate and compost to complete the composition.[†] However, compost should be used sparingly, as research from North Carolina State University showed that too much compost in the soil mixture allows excess nitrogen and phosphorous to leach from the green roof, thus negating some of the environmental purpose of the green roof.[‡] Other common growth media materials include expanded shale (brand name Haydite), expanded clay (brand name Arkalyte), volcanic pumice, scoria, and crushed clay roofing tiles.[§] A further important limitation linked to growth media is depth. In an extensive green roof system, growth media depth is typically 5 to 15 centimeters. Plants that develop large or deep root systems (many native plant species) will not be able to survive in shallow green roof growth media depths. The limited amount of inorganic nutrients in typical green roof media blends also may not be able to sustain plants with large nutritional requirements. A Michigan-based study showed that while *Sedum* species can thrive and survive in shallow media depths, native vegetation required a deeper media depth to survive.[¶]

Growth media composition will also directly affect plant growth and viability. As mentioned above, growth media can be composed of many different inorganic and organic entities. Inorganic components, such as Haydite, pumice, Arkalyte, and lava rock, provide proper drainage and moisture retention.

[*]Chicago Department of Environment, 2007.
[†]Sherman, 2005.
[‡]Sherman, 2005.
[§]Snodgrass and Snodgrass, 2006.
[¶]Monterusso et al., 2005.

Organic components, such as composted yard waste and composted pine bark, provide essential nutrients for plant sustainability. The ratio of inorganic to organic components is usually 70 to 80% and 20 to 30%, respectively. Note that the right choice can reduce fertilizer inputs to a green roof while still providing excellent conditions for growth and viability.

As an example, a Michigan-based experiment used heat-expanded slate (Haydite) as the inorganic substrate in growth media blends consisting of different inorganic component percentages (60 to 100%).* The remaining volume of growth media blend was composed of sand, Michigan peat, and aged compost (consisting of composted yard waste and aged poultry manure). This study utilized *Sedum* species and vegetation native to the Midwest. Plants grown in the green roof growth media composed of 100% Haydite generally showed the least amount of plant growth during the first 2 years of the study. This may be due to the lack of organic material to provide essential nutrients for plant growth. On the other hand, results clearly demonstrated that growing media composed of up to 80% Haydite (inorganic component) can be used without affecting plant health when using succulents, such as sedums, and can reduce the weight added to the roof of the building. The Michigan study also demonstrated that native plant species required more than 40% organic material in the growth media to survive and grow.

Many groups are searching for the next, best green roof growth media blend. Varying either the inorganic or organic component helps to get the best plant growth and environmental benefits. One recycled growth media that is being evaluated at SIUE consists of 80% spun-glass aggregate (a recycled product) and 20% composted pine bark. Initially, there was much interest in this growth media formulation because of the light weight of the spun glass and the possibility of a high storm water–retention capacity. However, in the green roof field trials this growth media remained saturated long after a rainfall event—even for long periods during drought. While promising in that it retains high moisture content, this glass-media blend stays too wet for sedums. (See Fig. 11.4.) The saturated root zone reduces plant growth and results in plant losses. Any other green roof

---

*Rowe et al., 2006.

Figure 11.4 Effect of growth media water-holding capacity on performance of *Sedum hybridum immergrunchen* [(*a*) 80% spun-glass aggregate and 20% composted pine bark and (*b*) 80% Arkalyte and 20% composted pine bark]. Project located at the SIUE green roof field site.

growth media formulation with really high moisture-holding capacity can lead to the same result—green roof failure.

Searching for the best growth media has led to many different green roof growth media mixes—the green roof trials at SIUE have used about 15 different growth media blends. Many commercial blends have a secret ingredient or two in the mix—something to outperform the other blends. Everyone is searching for the growth media blend that provides the best environmental benefit (storm water retention or thermal benefit) and nutritional requirements for the plants. In a comparison of media blends in modular green roof systems with four different inorganic components (Arkalyte, Haydite, lava, and pumice) blended 80:20 with composted pine bark, it was found that roof coverage by plants varied significantly in their first year on the roof. (See Fig. 11.5.) In fact, the roof coverage in the lightest-colored media (the pumice blend) was greater than all of the others at the end of the first season. Researchers have yet to determine whether temperature differences due to the "light color" of the inorganic media contributed to the increased roof coverage.

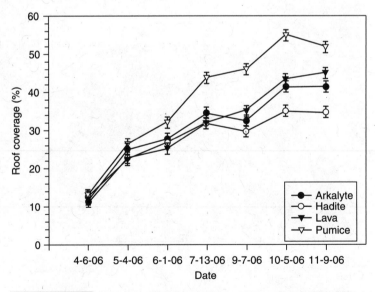

Figure 11.5 Plant roof coverage by sedums planted in different growth media during the 2006 growing season on the SIUE Engineering Building green roof (error bars ±1 standard error.). (Growth media inorganic:organic ratio was 80:20.) (Gibbs, 2008).

Besides the blended ingredients in the growth media, depth of the media mix matters greatly—not just for the weight issue. In one of the first field site green roof studies, built-in-place green roof systems of differing depths were designed. The media depth in this study (above the JDR drainage layer) was established at 5, 10, 15, and 20 centimeters in replicate green roof models. The growth media blend was 80% Arkalyte and 20% composted pine bark. *S. hybridum immergrunchen* was planted in September 2005 and plant growth, roof coverage, and storm water runoff have been monitored since that date. Plant survival is quite interesting—it was fully expected that the plants in the 5-centimeter growth media depth would not survive the extended droughts often experienced during summer in the St. Louis, Missouri, area. Turns out, that even following 35 days with no rain in the summer of 2006, there were still living plants in the 5-centimeter depth green roof models. (See Fig. 11.6.) However,

(a)

Figure 11.6   Effect of growth-media depth on performance of *Sedum hybridum immergrunchen* growing in 80% Arkalyte and 20% composted pine bark [(*a*) 5-cm depth and (*b*) 10-cm depth]. Project located at the SIUE green roof field site.

(b)

Figure 11.6   (*Continued*)

despite the plant survival in the 5-centimeter growth media, green roof industry collaborators will be the first to say that they cannot sell a green roof that is not substantially covered by green roof plants within the first 18 months. Fortunately, all of the other growth media depths in this study have suitable roof coverage (See Fig. 11.7.)—coverage that is more than adequate to perform as advertised.

Research—Does My Green Roof Work as Advertised?

Thirty-two built-in-place systems and four Green Roof Blocks were set up in completely randomized positions on four tables made from treated lumber at the Southern Illinois University Edwardsville (SIUE) campus in Edwardsville, Illinois. (See Fig. 11.8.) On one table, two glass rain gauges were mounted,

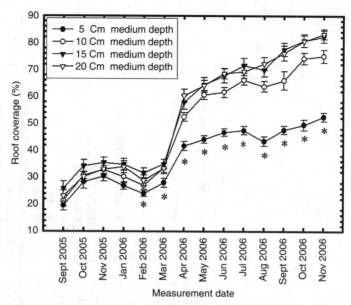

Figure 11.7 Roof coverage (%) by plants growing in built-in-place green roof systems at Southern Illinois University, Edwardsville, with 5, 10, 15, and 20 cm growth-media depths between September 5, 2005 to November 13, 2006 (error bars ±1 standard error.) (Forrester, 2008).

Figure 11.8 A green roof study located at the Southern Illinois University Edwardsville Environmental Sciences Field Site in Edwardsville, Illinois which included planted built-in-place green roof systems with 5, 10, 15, and 20 cm of growth media, unplanted built-in-place green roof systems with the same media depths, green roof blocks with 10 cm of growth media, and control roof models (standard EPDM roof surface).

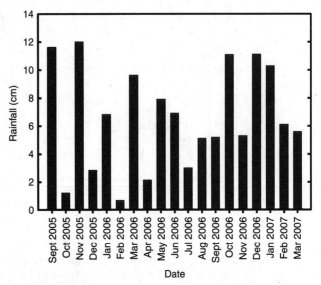

Figure 11.9 Total monthly rainfall collected in the green roof study at the Southern Illinois University Edwardsville Environmental Sciences field site for the study period of 9/05 to 4/07.

one on each end. Total rainfall for the study period was 124.48 centimeters between September 5, 2005 and April 1, 2007. (See Fig. 11.9.)

The percent of storm water retained by a green roof varies greatly from one geographic region to another due to rain frequency, intensity, and duration as well as the type of green roof, media selection, and plant material. In the first 18 months of the experiment at SIUE, storm water retention varied from 39% for green roof blocks to 53% for planted built-in-place systems containing 20 centimeters of growth media when compared to runoff from "control" EPDM roof surfaces. (See Fig. 11.10.) During the first 18 months of the study, much of the green roof surface was gradually covered by the *Sedum* species chosen. Now with mature plant coverage, the storm-water retention of the ongoing green roof storm water project has shown that in the 10-, 15-, and 20-centimeter growth media depths, more than 84% of storm water was retained between April 1, and November 13, 2007—an astounding success.*

---

*Woods, unpublished data.

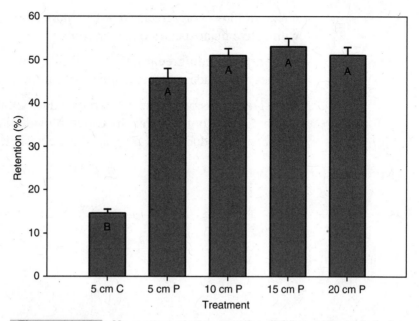

Figure 11.10   Mean storm water retention (%) for study period of 9/05 to 4/07 for green roof systems with 5, 10, 15, and 20 centimeters planted growth media depths. (C = control, P = planted, bars with same letter not significantly different at the $p < 0.05$ level. Error bars +1 standard error.) (Forrester, 2008).

This most recent data strongly suggests that green roofs do work as advertised—they reduce storm water runoff significantly.

Research and Resources

There are a lot of places one can look for quality green roof information and for up-to-date research information that can help get a green roof project started or to solve a green roof problem (other chapters of this book provide industry information for consultation about the green roof products themselves).

Here are some suggestions to get started:

www.green-siue.com—This site contains published research presentations and posters from the SIUE project.

www.greenroofs.com—The international portal for green roof information.

www.wildflower.org/greenroof—Dr. Mark Simmons works with native plant species on green roofs.

www.hrt.msu.edu/greenroof—Dr. Brad Rowe, his colleagues, and students provide green roof information.

www.greenroofs.org—The green roof industry trade organization, Green Roofs for Healthy Cities, provides information on green roof educational programs and current events.

Index